A TO Z HEALTH AND SAFETY HANDBOOK

A TO Z HEALTH AND SAFETY HANDBOOK

Tom O'Reilly

For Grace

First published in 2003 by Management Books 2000 Ltd
Forge House, Limes Road
Kemble, Cirencester
Gloucestershire, GL7 6AD, UK
Tel: 0044 (0) 1285 771441/2
Fax: 0044 (0) 1285 771055
E-mail: m.b.2000@virgin.net
Web: mb2000.com

Printed and bound in Great Britain by Biddles, Guildford

British Library Cataloguing in Publication Data is available
ISBN 1-85252-437-5

Introduction

Some years ago, having inherited a management responsibility for health and safety in a large company and with very little previous experience in the role, I was more than a little bewildered by the plethora of technical and jargon expressions I encountered almost daily. Health and safety is an eclectic mix of legal, occupational, health, scientific and management terms, ignorance of which may cause the unwary and unaware in this litigious age, to end up in a courtroom or industrial tribunal – or at the very least may sour relationships with staff or other departments.

Meetings with architects, facilities managers, contractors and trade union representatives were a minefield of technical and arcane terms. Discussion was difficult and deciphering their reports was a considerable headache. Training courses in health and safety were hastily arranged but, in the meantime, on looking around for a brief, short-hand guide to the problems, I found that there wasn't one. This book will, I hope, provide that short-hand guide and fill the gap between good common sense and what the law actually requires.

It is a sad fact that several hundred people lose their lives each year whilst at work. Nearly 160,000 are involved in non-fatal injuries, and over two million people are affected by ill health that has been caused by or is exacerbated by being at work. Annually, 25 million working days are lost through work-related accidents. There is a temptation to believe that 'it will never happen to me' or 'that sort of thing doesn't happen in our workplace'. Wrong!

Taking care of yourself and other people at work is vital. If you are an employer or an employee, you have many statutory duties and responsibilities as well as many simple, common-sense things that you can do to prevent injury or damage to health. By taking sensible precautions and helping to create a safe working environment you can prevent people from being harmed.

Such is the importance of this, that there are many laws, regulations and codes of practice to enforce or, at the very least, encourage good habits and working behaviours. Many of these laws are shown in the book at the relevant topic. There are also many court cases which have had a very significant bearing on health and safety – some of these are also listed.

It is important to recognise that health and safety laws and practices apply to everyone, whether in large or small businesses, or self-employed.

Inspectors from the Health and Safety Executive (HSE) or the local authority have the responsibility of administering these laws and can enforce them when things go wrong.

All businesses – even the one-man band – have a responsibility to take the necessary steps to ensure that no one's health or safety is jeopardised and that the correct policies, practices and precautions are understood and carried out by everyone.

This book takes a comprehensive and careful look at the whole range of potential hazards that may face a worker. It is not intended to be alarmist, but everything included is there for a purpose. It works at two levels – it provides a quick definition of the newly-encountered phrase or term and where possible, points the reader towards the primary source where more detailed help and information can be found. The book does not pretend to provide all the answers to all the problems – that would take a whole range of books – but it will set you off in the right direction.

Work safely and live healthily – good luck.

Tom O'Reilly
October 2003

Acknowledgements

It would not be possible to produce a book such as this without the help of those organisations which are involved in producing guidance and literature for health and safety practitioners. I would therefore like to thank HMSO, HSE Books and NCEC for their willingness to allow me to use some of their material and ideas.

Abbreviations and Acronyms

The following is a list of abbreviations and acronyms which appear in this A to Z Handbook

ACAS	Advisory Conciliation and Arbitration Service
ACOP	Approved Code of Practice
ACTS	Advisory Committee on Toxic Substances
ADR	European Agreement on International Carriage of Dangerous Goods by Road
AIDS	Acquired Immunodeficiency Syndrome
APC	Air Pollution Control
BATNEEC	Best available techniques not entailing excessive cost
BI 510	Accident Book (Approved)
BLEVE	Boiling liquid expanding vapour explosion
BNFL	British Nuclear Fuels
BPEO	Best practical environmental option
BS	British Standard
BSC	British Safety Council
BSI	British Standards Institute
CATNAP	Cheapest available technique not attracting prosecution
CDM	Construction Design and Management Regulations
CE	Commission Européene
CEN	Comité Européen de Normalisation
CENELEC	Comité Européen de Normalisation Electrotechnique
CFC	Chlorofluorocarbon
CHIP	Chemicals (Hazard Information and Packaging for Supply) Regulations
CITB	Construction Industry Training Board
COMAH	Control of Major Accident Hazards Regulations
COSHH	Control of Substances Hazardous to Health Regulations
dB	Decibel
dB(A)	'A' weighted decibel – Occupational noise unit of measurement
DEFRA	Department for Environment, Food and Rural Affairs
DTI	Department of Trade and Industry
DDA	Disability Discrimination Act
DSE	Display Screen Equipment Regulations
EAT	Employment Appeals Tribunal

EC	European Commission
EEC	European Economic Community
EHO	Environmental Health Officer
EINECS	European Inventory of Existing Chemical Substances
ELCB	Earth leakage circuit breaker
ELINCS	European List of Notified Chemical Substances
EMAS	Employment Medical Advisory Service
EN	European Normalisation prefix
EU	European Union
FLT	Fork lift Truck
FMEA	Failure Mode and Effect Analysis
HACCP	Hazard analysis and critical control points
HAZCHEM	Hazardous chemicals code
HAZOP	Hazard and operability study
HSC	Health and Safety Commission
HMIP	Her Majesty's Inspectorate of Pollution
HMSO	Her Majesty's Stationery Office
HFL	Highly flammable liquid
HIV	Human Immunodeficiency Virus
HSE	Health and Safety Executive
HSWA	Health and Safety at Work etc Act
Hz	Hertz
IEE	Institute of Electrical Engineers
IOSH	Institute of Occupational Safety and Health
IPC	Integrated Pollution Control
ISO	International Standards Organisation
IUPAC	International Union of Pure and Applied Chemistry
LASER	Light amplification by stimulated emission of radiation
LC_{50}	Lethal concentration
LD_{50}	Lethal dose
LEV	Local exhaust ventilation
LOLER	Lifting Operations and Lifting Equipment Regulations
LPG	Liquid Petroleum Gas
LEV	Local exhaust ventilation
MHSW	Management of Health and Safety at Work Regulations
MHOR	Manual Handling Operations Regulations 1992
MEL	Maximum exposure limit
MIIRSM	Member of the International Institute of Risk and Safety Management
NEBOSH	National Examination Board in Occupational Safety and Health
NICEIC	National Inspection Council for Electrical Installation Contracting

NRPB	National Radiation Protection Board
OES	Occupational Exposure Standard
pH	Measure of acidity/alkalinity
PPE	Personal Protective Equipment
PUWER	Provision and Use of Work Equipment Regulations
PVC	Polyvinyl chloride
RCD	Residual current device
REM	Roentgen Equivalent Man
RIDDOR	Reporting of Injuries Diseases and Dangerous Occurrence Regulations
RoSPA	Royal Society for the Prevention of Accidents
RPE	Respiratory Protective Equipment
RSI	Repetitive Strain Injury
STEL	Short Term Exposure Limit
SWL	Safe working load
TLV	Threshold limit value
TREMCARD	Transport emergency card
TWA	Time weighted average
WRULD	Work-related upper limb disorder
WTR	Working Time Regulations

Notes

🕮 This symbol is use to indicate a court case that has particular significance to the topic being considered.

❑ This symbol is used to indicate a relevant Act of Parliament or other piece of legislation. You are advised to refer to these items for a fuller description of the matter.

Words in *italics* have a separate entry in the text.

➔ This symbol indicates a further useful reference

• ABRASIVE WHEEL

A wheel, cylinder, disc or point having abrasive particles and intended to be power driven, used for grinding, cutting or sharpening

Improper or untrained use often leads to accident or injury. References may still be seen to the Abrasive Wheels Regulations but these have now been replaced by the Provision and Use of Work Equipment Regulations.

- ❑ ***Abrasive Wheels Regulations 1970***
- ❑ ***Provision and Use of Work Equipment Regulations 1998***

• ABSOLUTE

In legal terms, this means that where there are no qualifying words or phrases accompanying the regulation, the regulation or requirement must be obeyed

For example, 'It shall be the duty of every employee while at work to take reasonable care for the health and safety of himself and of other persons who may be affected by his acts or omissions at work.' The duty here is absolute and must be complied with. ➔ *liability*

- ❑ ***Health and Safety at Work etc Act 1974***

• ABSORPTION

A process in which one substance penetrates into the interior of another solid or liquid substance

Absorption is also classified as one of the routes of entry through which toxic substances may enter the human body. A substance can obviously enter the body through cuts and abrasions but may also be absorbed by contact through the cells of the skin or even hair follicles

• ACAS

Acronym for the Advisory, Conciliation and Arbitration Service

The Service mediates, in industrial relations problems, between employees and employers. It has published a *Code of Practice* on disciplinary and grievance matters

• ACCELERANT

A substance used to start and increase the rate of fire, e.g. *flammable liquid*

• ACCESS

Access, in a health and safety context, does not simply mean entry and exit to and from buildings, but access to the specific workplace which could be a room, corridor, road, platform, scaffold, etc

The employer must, so far as is reasonably practicable, provide and maintain the means of access to and egress from any place of work under his control in a condition that is safe and without risks to health.

❑ ***Health and Safety at Work etc Act 1974***

• ACCIDENT

An accident, according to the HSE is any undesired circumstances which give rise to ill-health or injury; damage to property, plant, products or the environment; production losses, or increased liabilities

➔ *Incident*

An accident, according to *RIDDOR*, includes an act of non-consensual physical violence done to a person at work and an act of suicide which occurs on, or in the course of the operation of, a relevant transport system

❑ ***Reporting of Injuries, Diseases and Dangerous Occurrences 1995***

• ACCIDENT BOOK

An approved document required by the HSWA 1974

An employer with ten or more employees must keep an accident book in which details of accidents, injuries, near-miss incidents and work-related stress symptoms or breakdowns are recorded. The approved book is a BI 150 obtainable from HMSO. From January 2004, all personal information entered in an accident book must be kept confidential to comply with the Data Protection AcT 1998.

● ACCIDENT CAUSES

Accident causes fall into two categories: a) the immediate event or circumstance that caused the accident, and b) the root cause – the sequence of events or circumstances which led to the immediate event that caused the accident

➔ *Domino Theory*

● ACCIDENT COSTS

Accident costs are divided between insured and uninsured costs:

Insured Costs

Employer's liability claims
Public liability claims
Building/vehicle damage
Business interruption
Product liability

Uninsured Costs

Insurance excess
Indemnity limit excess
Legal costs
Sick pay
Repairs
Product loss
Plant/tool/equipment damage
Site clearance
Production delays
Overtime working
Investigation cost
Bad publicity
Re-placement/retraining staff
Management time loss
Lost business opportunity

● ACCIDENT STATISTICS

In compiling accident statistics the following standard indices are used:

Frequency rate $$\frac{\text{total number of accidents} \times 1{,}000{,}000}{\text{total number of man hours worked}}$$

Incidence rate $$\frac{\text{total number of accidents} \times 1000}{\text{average number of persons exposed}}$$

Severity rate	total number of days lost x 1000 / total man hours worked
Duration rate	number of man hours worked / total number of accidents
Mean duration rate	total number of days lost / total number of accidents

• ACCIDENT THEORIES

The following accident theories are cited in many accident studies:

Heinrich, in a 1950s study, estimated that in a group of 330 accidents there would be:

1	major injury
29	minor injuries
300	no injury accidents

Bird, in a 1969 study – an analysis of over 1.75 million accidents – estimated the following ratios:

1	serious or disabling injury
10	minor injuries
30	property damage accidents
600	no visible injury or damage

Tye and Pearson, in a 1974/75 research study, based on just under a million accidents, produced the following estimated ratios:

1	fatal or serious injury
3	minor injuries
50	injuries requiring first-aid
80	property damage accidents
400	no injury/damage incidents or near misses

HSE Accident Prevention Advisory Unit, in a 1997 study, suggested the following ratio:

1	major or over 3-day lost-time injury
7	minor injuries
189	non-injury accidents

All the studies, although based on different figures and ranging over a wide time-scale, indicate that there is a consistent relationship between the different kinds of events.

• ACCOMMODATION FOR CLOTHING

Suitable and sufficient accommodation shall be provided at work for any person's own clothing which is not worn during working

hours and for special clothing worn at work but which is not taken home

❑ ***Workplace (Health, Safety and Welfare) Regulations 1992***

• ACOUSTIC SIGNAL

A coded sound signal which is released and transmitted by a device designed for that purpose, without the use of a human or artificial voice

❑ ***Health and Safety (Safety Signs and Signals) Regulations 1996***

• ACQUIRED IMMUNODEFICIENCY SYNDROME (AIDS)

AIDS is a disease caused by the *HIV* virus transmitted between individuals through sexual contact, through infected blood transfusions, by contact with the body fluids of an infected person, and by the use of unclean hypodermic needles

• ACTINOID

The group name for the elements numbered 89 – 103. Actinoids contained in nuclear waste continue to emit radiation for long periods, giving rise to storage and disposal problems

• ACTION LEVEL

The level of exposure to a health hazard at which an employer must take preventive action to protect his employees by eliminating the hazard or reducing the exposure to it

❑ ***Control of Asbestos at Work Regulations 1987 and Noise at Work Regulations 1989***

• ACTS

The acronym for Advisory Committee on Toxic Substances which regulates *MEL*s and *OES*s

• ACTUS REUS

The Latin term which describes all the elements in the definition of a crime with the exception of *mens rea*

• ADR

The abbreviation for the European Agreement concerning the International Carriage of Dangerous Goods by Road

An ADR mark is a mark specified in the Agreement to indicate that packagings which carry it correspond with a design type which has been successfully tested in accordance with ADR and comply with its manufacturing provisions.

❑ ***Carriage of Dangerous Goods (Classification, Packaging and Labelling) and Use of Transportable Pressure Regulations 1996 (as amended 1999)***

● ADSORPTION

The power of a solid substance to attract a vapour or solution to its surface

● ADULT WORKER

A worker who has attained the age of 18

❑ ***Working Time Regulations 1998***

● AEROBIC

Aerobic means requiring the presence of oxygen for living, functioning and growth

● AEROSOL

A system of finely distributed particles in a gas or in the air

Dust is technically an aerosol although these days the term is more commonly applied to pressurised spray cans containing paint, hair lacquer, etc.

● AEROSOL DISPENSER

An article which consists of a non-refillable receptacle containing a gas compressed, liquefied or dissolved under pressure, with or without liquid paste or powder and fitted with a release device allowing the contents to be ejected as solid or liquid particles in suspension in a gas, as a foam, paste or powder or in a liquid or gaseous state

❑ ***Carriage of Dangerous Goods (Classification, Packaging and Labelling) and Use of Transportable Pressure Receptacles Regulations 1996 (as amended 1999)***

● AETIOLOGY

The study of the causation of diseases

● AGRICULTURE

Agriculture includes horticulture, fruit growing, seed growing,

dairy farming, livestock breeding and keeping, forestry, the use of land as grazing land, meadow land, osier land or nursery grounds or for market gardening and the preparation of land for agricultural purposes

❑ *Environmental Protection Act 1990*

• AGRICULTURAL WASTES

Agricultural wastes are divided into four categories – silage, liquors, straw and chemical wastes, and containers which have been used for sheep dips, pesticides and fertilisers

❑ *Environmental Protection Act 1990*

• AIDS

The acronym for acquired immunodeficiency syndrome

• ALGAE

Any of a large group of simple plants containing chlorophyll but without stems, roots or leaves

• ALGORITHM

A defined process or set of rules that leads to a logical conclusion – used in calculations or problem-solving operations

• ALIMENTARY CANAL

The passage by which food passes through the body

• ALTERNATING CURRENT

An electric charge which reverses its flow periodically

➔ direct current

• ALLERGEN

A substance, such as pollens, dust, and food products, which irritates the affected part of the body because of the inability of the body's immune system to cope with it

• ALLERGY

A reaction of the body's immune system to an *allergen*

• ALPHA

In radiation, α or alpha radiation is weak and an α particle would be unlikely to penetrate the skin

Penetration can be prevented by a layer of skin or a sheet of paper. (see ionization). The main danger from alpha particles is where they are ingested and can damage internal organs which have no skin protection.

• ALVEOLUS

A tiny balloon-like sac arranged in millions of bunches inside the lung

The walls of the alveoli contain blood vessels through which an exchange of oxygen and carbon dioxide in the blood takes place.

• AMBIENT TEMPERATURE

The temperature of the surrounding environment

• AMINO ACID

A basic chemical substance from which proteins are synthesised by the body – essential to all forms of life

• AMOSITE

➔ asbestos

• AMPHIBOLE

One of a group of silicate minerals having a crystal structure containing aluminium, calcium, iron, magnesium and sodium ions

• ANAEROBIC

Capable of living and growing without the use of oxygen

• ANGLE OF REPOSE

The angle beyond which excavated materials will tend to slide

This varies enormously from rock (90°) to clay which can be as low as 5° when wet. Water saturation greatly reduces the angle of repose of materials such as clay. A stable secure excavation that needs no support when dry can collapse easily when wet.

• ANTHRAX

One of the most virulent diseases of animals

Cattle, sheep, goats, horses, etc, i.e. grazing animals, contract the disease by grazing on infected pasture. The disease can be transmitted to humans by handling of wool, hair, hides, bones or carcases of infected animals. It is a reportable disease under *RIDDOR* .

❑ ***Reporting of Injuries, Diseases and Dangerous Occurrences Regulations 1995***

• ANTHROPOMETRICS

A system of measuring and recording the sizes and proportions of the human body to provide data which can be used, amongst other purposes, in the design of furniture, equipment and machinery

• APC

The abbreviation for air pollution control which is administered and enforced by local authorities

It is concerned only with emissions into the atmosphere which may be considered to have a pollutant effect on the environment. ➔ *IPC*

❑ ***Environmental Protection Act 1990***

• APPEALS

A system of challenging decisions made by criminal and civil courts

• APPELLANT

A person bringing an appeal in a civil or criminal action

• APPOINTED DOCTOR

A registered medical practitioner who is for the time being appointed in writing by the HSE for the purposes of specified regulations

❑ ***Control of Lead at Work Regulations 1998***
❑ ***Ionising Radiations Regulations 1999***

• APPOINTED PERSON

An unqualified person who can, nevertheless, take charge of the first-aid box and be available at all times to summon medical help

This is a requirement which applies to even the smallest workplace.

➔ *First-aider* and *suitable person*

❑ ***Health and Safety (First Aid) Regulations 1981***

• APPROVED

Approved means approved for the time being in writing for the purposes of regulations by the HSE or HSC

• APPROVED CARRIAGE LIST

A list of dangerous goods (not including radioactive materials or explosives) which are approved by the HSC for carriage by road and rail

➔ *risk* and *safety phrases*

❑ ***Carriage of Dangerous Goods (Classification, Packaging and Labelling) and Use of Transportable Pressure Receptacles Regulations 1996***

• APPROVED CODE OF PRACTICE

Quasi-legal documents issued by the *Health and Safety Commission*

They supplement and explain the law and give guidance on compliance with the legislation. The HSC can issue and approve codes of practice under Section 16 of the HSAW Act and can approve other codes and standards issued by other organisations such as the British Standards Institute. Non-compliance does not constitute a breach of the law but ACOPs are used as evidence of failure to do all that was reasonably practicable to comply with the law. If a similar or better standard of compliance can be demonstrated there is no need to comply with the advice contained in the Code of Practice. There are currently three series of codes – HS(R); COP; and L.

➔ *Appendix 2* for some examples of each.

❑ ***Health and Safety at Work etc Act 1974***

• APPROVED DOSIMETRY SERVICE

A dosimetry service approved in accordance with the Regulations

❑ ***Ionising Radiations Regulations 1985***

• APPROVED PERSON

A person approved by a competent authority for the purpose of carrying out certain functions in connection with the examination, testing and certification of equipment and vehicles

❑ ***Pressure Systems and Transportable Gas Containers Regulations 1989***

• APPROVED SUPPLY LIST

A list approved by the *HSC* containing information regarding the classification and labelling of dangerous substances and preparations

➔ *risk* and *safety phrases*

❑ ***Chemicals (Hazard Information and Packaging for Supply) Regulations 1996***

• ARM'S REACH

A zone of accessibility to touch extending from any point on a surface where a person usually stands or sits or moves about, to the limits of reach with the hands in any direction without assistance

• ARTICLE 100A

An article of the Single European Act 1986 which enabled agreement to be achieved, on the approximation of laws within the Member States of the EU, by *qualified majority vote*

❑ ***Single European Act 1986***

• ARTICLE 118A

Article 118A is an article of the Single European Act 1986 which enabled agreement to be reached, on health and safety directives within the Member States of the EU, by *qualified majority vote*

❑ ***Single European Act 1986***

• ASBESTOS

A natural mineral fibre obtained by means of a mining or quarrying process

Originally used extensively in construction for its non-flammability and thermal insulation. There are a number of asbestos types:

Amosite, crocidolite, chrysotile, fibrous actinolite, fibrous anthophyllite, fibrous tremolite and any mixture containing any of those minerals. The three main types are:

Amosite known as brown asbestos
Crocidolite known as blue asbestos
Chrysotile known as white asbestos.

Exposure to any of these may lead to *asbestosis* (q.v.). Action levels and control limits for exposure to asbestos are specified by statute.

❑ ***Control of Asbestos at Work Regulations 1987***

• ASBESTOSIS

Asbestosis is a form of fibrosis of the lungs brought about by inhalation of asbestos fibres causing extreme breathlessness and which may cause cancer of the lung

• ASSESSMENT

A legal obligation placed on all employers to evaluate the hazards and risks to the health and safety of their employees.

➔ *risk assessment*

• ATTENUATION

A reduction in strength or value

In health and safety it is most commonly used to describe the reduction or loss of sound.

• ASPHYXIA

Unconsciousness brought about by deprivation of oxygen to the lungs

• AUDIOMETER

A device for indicating hearing acuity to determine the threshold of hearing and to measure hearing loss

• AUDIOMETRY

The measurement of the perception of hearing at various sound frequencies

➔ *noise levels*

• AUDIT

➔ *safety audit*

• AUTOMATIC FIRE DETECTION

A system designed to detect changes in the protected area indicating the development of a fire situation

The systems operate when the invisible products of combustion are released into the area; or when smoke is being produced; or when there is either a rapid rise in temperature or the temperature has risen to a pre-determined figure. ➔ *heat* and *smoke detectors*

• BANKSMAN

A person who communicates with a crane or derrick driver by means of hand signals in order to direct lifting operations

➔ *safety signals* and *hand signal*

❑ ***Health and Safety (Safety Signs and Signals) Regulations 1996***

• BASE PLATE

A square metal plate about 150 x 150 mm which is placed on a sole plate under a scaffold standard to provide level support

➔ *scaffold, sole plate, standard*

• BENZENE

A hydrocarbon which is toxic, colourless, aromatic and flammable. It is acutely narcotic and is an irritant to the skin

• BATNEEC

Best available techniques not entailing excessive costs

Where an industrial process has been authorised, in order to minimise pollution to the environment, the best available techniques not entailing excessive costs must be applied to the process.

❑ ***Environmental Protection Act 1990***

• BETA

The beta (β) ray, in radiation, has stronger penetrative power than the alpha particle

➔ ***ionization***

• BEYOND A REASONABLE DOUBT

The standard of proof required to prove guilt in criminal law cases

BINDING PRECEDENT

In court procedure, decisions on similar circumstances made by higher courts are said to be binding on inferior courts, i.e. lower courts must follow the precedent thus established

BI-METALLIC STRIP

A thin strip of metal in which steel and brass are bonded together

When subjected to heat, because of the differing rates of expansion of the metals, the strip will bend. This factor is used in many thermostats and temperature control devices but is particularly used in heat detectors where the expansion of the strip is used to complete an electrical circuit thus sounding an alarm.

BIODEGRADABLE

A term applied to material (usually waste products) which are capable of decomposing (degrading or breaking down) in the environment by natural biological processes

❑ ***Control of Substances Hazardous to Health Regulations 1999***

BIOHAZARD SIGN

A sign which must be displayed where there is a risk of exposure to a biological agent

❑ ***Control of Substances Hazardous to Health Regulations 1999***

BIOLOGICAL AGENT

A biological agent is defined as any micro-organism (bacteria, fungi, viruses, and microscopic parasites), cell cultures, human endoparasites which might cause infection, allergy, toxicity or otherwise create a hazard to health

❑ ***Control of Substances Hazardous to Health Regulations 1999***

BIOLOGICAL MONITORING

The measurement of a person's blood-lead concentration or urinary lead concentration in accordance in either case with the method known as atomic absorption spectroscopy

❑ ***Control of Lead at Work Regulations 1998***

● BLEVE

The acronym for boiling liquid expanding vapour explosion

An example of this is where a heavy fuel oil contained in a tank catches fire. The heat can extend down through the oil and if there is water in the tank bottom the hot oil coming into contact with it will turn the water into steam causing an explosion.

● BOATSWAIN'S CHAIR (Bosun's Chair)

A chair suspended by ropes or chains in which a person can sit in order to reach parts of a structure which cannot be easily reached from a working platform

➔ *personal suspension equipment*

❑ ***Construction (Health, Safety and Welfare) Regulations 1996***

● BOTULISM

Acute poisoning of the system caused by ingesting toxins from contaminated food

● BPEO

The acronym for best practical environmental option – a term used in connection with control of environmental pollution

In order to achieve a given set of objectives which might impact on the environment, the BPEO must be the best option for the provision of the greatest advantage, or the least damage, to the environment at an acceptable cost.

❑ ***Environmental Protection Act 1990***

● BREACH OF DUTY

Breach of Statutory Duty is the breaking of an obligation imposed by a statute

● BREMSSTRAHLUNG

Electromagnetic radiations produced by the slowing down of a ß particle

As with a ß particle, they have considerable powers of penetration.

● BRITISH STANDARD

A British Standard is a standard approved by the British

Standards Institute which sets national legal standards for manufacturing industries in the UK.

➔ *CEN, CENELEC*

• BNFL

The abbreviation for British Nuclear Fuels Limited, a company involved in the manufacture, processing, and waste management of nuclear products

• BSC

The abbreviation for the British Safety Council, an organisation concerned with all aspects of health and safety and which runs training courses in safety awareness and safety management. It provides training and tuition leading to a diploma in safety management and environmental management

• BS

The abbreviation for British Standard

Any product, machinery or piece of equipment which has been awarded a British Standard has achieved a standard which is in compliance with recognised and accepted safety procedures. ➔ *EN*

• BSI

The abbreviation for the British Standards Institute which is the leading organisation which sets standards for manufacturing industries in the UK

• BS 8800

A quality control standard for the management of health and safety

❑ ***Guide to Occupational Health and Safety Management Systems***

• BS EN

BS EN is a British Standard which also complies with an associated European Standard

➔ *EN*

• BS EN ISO 9000

BS EN ISO 9000 is an International Quality Standard for Health and Safety Management

• BS EN ISO 14000; 14001; 14004; 14010; 14011; 14012;14040

These are a series of standards concerned with environmental systems covering management, auditing and assessment techniques

• BRUCELLOSIS

A fever occurring in persons who are involved with handling live cattle or pigs or their carcasses. It is a reportable disease under *RIDDOR*

❑ Reporting of Injuries, Diseases and Dangerous Occurrences Regulations 1995

• BUILDING REGULATIONS

Regulations governing the design and construction of buildings to ensure that they are constructed properly and are safe for occupation

• BURDEN OF PROOF

In English law, the burden of proof is said to rest on the person alleging the truth of a particular issue or statement

In criminal law, generally, the prosecution have to prove, beyond a reasonable doubt, that the defendant committed the alleged offence. In civil law the standard of proof is based on the balance of probabilities – a much lower standard.

In health and safety offences, the burden of proof switches from the prosecution to the defendant, i.e., once it has been established that an offence under the HSWA has taken place, the accused has to prove his innocence.

• BURIED SERVICES

Buried services are utilities such as gas, water, electricity and telecommunications pipes and cables which are below ground and present a hazard when excavation is taking place

• BURSTING DISC

A diaphragm in a pressure vessel or pipe-work designed to burst at a pre-determined pressure

• BUND

A raised, enclosed area around liquid storage tanks to contain spillage should the contents of the tanks escape

The bund capacity should be at least 110% of the capacity of the storage tank.

❑ ***Highly Flammable Liquids and Liquefied Petroleum Gases Regulations 1972***

• BUSBAR

A bare electrical conductor forming a common junction between a number of circuits

• BYSSINOSIS

A lung condition occurring in persons exposed to cotton dust.

It is characterised by asthmatic attacks and can result in *emphysema*. It is a reportable disease under *RIDDOR* .

❑ ***Reporting of Injuries, Diseases and Dangerous Occurrences Regulations 1995***

• CAISSON

A construction, usually round, built partly or wholly above ground (or water) usually by digging out the earth inside it

Caissons are often used as supports or piers in bridge-building.

• CALENDAR YEAR

Calendar year means the period of twelve months beginning with 1 January in any year

❑ ***Working Time Regulations 1998***

• CAMPYLOBACTER

A bacterium which causes severe food poisoning in humans

• CANCER

A group of diseases caused by the unrestrained growth of cells in the human body, which may be benign or malignant

• CARCINOGEN

Generally, a substance which promotes the growth of a cancer

More specifically, any substance classified under the Chemical (Hazard Information and Packaging for Supply) Regulations as requiring to be labelled with the risk phrases 'may cause cancer' or 'may cause cancer by inhalation' whether or not the substance requires to be classified anyway under COSHH

❑ ***Chemical (Hazard Information and Packaging for Supply) Regulations 1996***

• CARCINOGENOSIS

The development of a cancer caused by the action of viruses, chemicals, radiation, etc

• CARPAL TUNNEL SYNDROME

A painful condition of the fingers caused by undue pressure on the median nerve where it passes through the carpal tunnel at the front of the wrist

• CASE LAW

Decisions of the higher courts arrived at, after hearing a case, which then become part of the law

➔ ***precedent*** **and** ***health and safety case law***

• CASE STATED

Case stated is a method of appealing after a criminal case has been heard by magistrates

The defence or prosecution may make the appeal and the magistrates set out a statement of the relevant facts of the case (the 'case stated') which is then submitted to a higher court, normally the Queen's Bench Divisional Court, for the opinion of the judge on a particular point of law.

• CATALYST

A substance which controls the rates of chemical reactions

• CRT

The acronym for cathode ray tube – a vacuum tube used in televisions and computer monitors

• CATNAP

An acronym for cheapest available techniques not attracting prosecution

• CATEGORY OF DANGER

Category of danger means, in relation to a substance or preparation dangerous for supply, one of the categories of danger specified in column 1 of Part 1 of Schedule 1 of the Regulations

➔ ***classification of chemicals***

Schedule 1

Column 1 Category of Danger	Column 2 Property	Column 3 Symbol-letter
PHYSICO-CHEMICAL PROPERTIES		
Explosive	Solid. liquid, pasty or gelatinous substances and preparations which may react exothermically without atmospheric oxygen thereby quickly evolving gases, and which under defined test conditions detonate, quickly deflagrate or upon heating explode when partially confined.	E
Oxidising	Substances and preparations which give rise to a highly exothermic reaction in contact with other substances, particularly flammable substances	O
Extremely flammable	Liquid substances and preparations having an extremely low flashpoint and a low boiling point and gaseous substances and preparations which are flammable in contact with air at ambient temperature and pressure.	F +
Highly flammable	The following substances and preparations, namely- (a) substances and preparations which may become hot and finally catch fire in contact with air at ambient temperature without any application of energy. (b) solid substances and preparations which may readily catch fire after brief contact with a source of ignition and which continue to burn or to be consumed after removal of the source of ignition. (c) liquid substances and preparations having a very low flash point (d) substances and preparations which, in contact with water or damp air, evolve highly flammable gases in dangerous quantities	F
Flammable	Liquid substances and preparations having a low flash point	none
HEALTH EFFECTS		
Very toxic	Substances and preparations which in very	T +

	low quantities cause death or acute or chronic damage to health when inhaled, swallowed or absorbed via the skin	
Toxic	Substances and preparations which in low quantities cause death or acute or chronic damage to health when inhaled, swallowed or absorbed via the skin	T
Harmful	Substances and preparations which may cause death or acute or chronic damage to health when inhaled, swallowed or absorbed via the skin	Xn
Corrosive	Substances and preparations which may, on contact with living tissues, destroy them	C
Irritant	Non-corrosive substances and preparations which, through immediate, prolonged or repeated contact with the skin or mucous membrane, may cause inflammation	Xi
Sensitising	Substances and preparations which, if they are inhaled or if they penetrate the skin, are capable of eliciting a reaction by hypersensitisation such that on further exposure to the substance or preparation, characteristic adverse effects are produced	
Sensitising by inhalation		Xn
Sensitising by skin contact		Xi
Carcinogenic	Substances and preparations which, if they are inhaled or ingested or if they penetrate the skin, may induce cancer or increase its incidence	
Category 1		T
Category 2		T
Category 3		Xn
Mutagenic	Substances and preparations which, if they are inhaled or ingested or if they penetrate the skin, may induce heritable genetic defects or increase their incidence	
Category 1		T
Category 2		T
Category 3		Xn
Toxic for reproduction	Substances and preparations which, if they are inhaled or ingested or if they penetrate the skin, may produce or increase the incidence of	

	non-heritable adverse effects in the progeny and/or an impairment of male or female reproduction functions or capacity	
Category 1		T
Category 2		T
Category 3		Xn
Dangerous for the Environment	Substances which, were they to enter into the environment, would present or might present an immediate or delayed danger for one or more components of the environment	N

❑ ***Chemicals (Hazard Information and Packaging for Supply) Regulations 1994***

• CDM

CDM is the abbreviation for the Construction (Design and Management) Regulations which govern the minimum health and safety requirements at temporary or mobile construction sites

Under the Regulations construction work is notifiable to the HSE if:

- it is scheduled to last for more than 30 days, or
- will involve more than 500 man-days, or
- includes any demolition work, or
- involves 5 or more workers on site at any one time.

❑ ***Construction (Design and Management) Regulations 1994***

• CE MARKING

CE Marking is the conformity marking affixed to equipment or protective systems signifying that the equipment conforms to approved European standards

The marking consists of the letters CE followed by an identification number.

• CEN

Comité Européen de Normalisation

The European Committee for Standardisation together with *CENELEC* are the bodies responsible for setting European standards which are acceptable to and agreed with each of the Member States of the EC.

• CENELEC

Comité Européen de Normalisation Electrotechnique – the European Committee for Electrotechnical Standardisation

➔ *CEN*

• CENTRE-TAPPED TO EARTH

110 volt centre-tapped to earth is an electrical safety system used on construction sites where a 110v transformer has been used to reduce a 240v electricity supply to 110v

The secondary winding of the transformer is centre-tapped to earth – this puts the line conductor at +55v and the neutral at -55v thereby reducing the maximum shock voltage in the event of an earthing problem to 55v.

• CHILD

A person who is not over compulsory school age

➔ *young person*

❑ ***Management of Health and Safety at Work Regulations 1999***

• CHIP

The acronym for Chemicals Hazard Information and Packaging – the regulations governing the classification, packaging and labelling of dangerous substances for commercial and industrial supply

Hazardous chemicals must be properly classified, packaged and labelled before supply to industry or consumers. Labels must include the name and address of the supplier, the name of the substance, the hazard symbol indicating the degree of danger, appropriate risk and *safety phrases*, and the EEC number . ➔ *classification of chemicals*

❑ ***Chemicals (Hazard Information and Packaging for Supply) Regulations 1994***

• CFC

The abbreviation for Chlorofluorocarbons

These are substances used as coolants in refrigerators and air-conditioners and also found in some solvents. On release into the atmosphere, they cause damage to the ozone layer. ➔ *Montreal Protocol*

CHLOROPHYLL

A green pigment present in plant cells giving plants their distinctive green colour

CIRCADIAN RHYTHM

The way in which, during a 24 hour cycle, the physiological and psychological characteristics of the human body operate in a rhythmic manner

A number of studies have found that the alteration of the circadian rhythms of shift workers can affect their work performance.

CIRCUIT

Circuit is the path taken through cabling by electric current to supply electrical equipment

CIVIL COURTS

Civil courts are courts whose function is to deal with aspects of civil law rather than criminal law

Courts having civil jurisdiction are the High Court and County Court. The High Court has three Divisions – Queen's Bench, Chancery, and Family.

CIVIL LAW

Civil law covers non-criminal matters such as probate, divorce, tort, breach of contract, etc

CIVIL LIABILITY

➔ *liability*

CLASSIFICATION OF CHEMICALS FOR SUPPLY

A substance or preparation which is dangerous for supply must be classified by placing it into one or more of the categories of danger specified in Schedule 1

The categories are in three parts:

- Substances and preparations dangerous because of their physical or chemical properties – explosive; oxidising; extremely flammable, highly flammable, flammable
- Substances and preparations dangerous because of their effects on health: very toxic; toxic; harmful; corrosive; irritant, carcinogenic, mutagenic, teratogenic
- Substances dangerous for the environment

Substances and preparations are classified if they are contained in the Approved Supply List or have been notified in accordance with the Notification of New Substances Regulations 1993. If neither applies, a substance or preparation must be classified by placing it into one of the above-mentioned categories of danger and assigning to it an appropriate risk phrase. ➔ *categories of danger*

❑ ***Chemicals (Hazard Information and Packaging for Supply) Regulations 1994***

• CLASSIFICATION OF FIRE

Fires are classified according to the fuel involved and the method of extinguishing the fire:

Class A Fires involving solids, wood, paper, coal etc
Extinguishing Agent Water

Class B Fires involving liquids
Extinguishing Agent Foam, carbon dioxide and dry powder

Class C Fires involving gases
Extinguishing Agent Foam

Class D Fires involving metals
Extinguishing Agent Dry powder

Electrical Fires involving electrical equipment
Extinguishing Agent Carbon dioxide or dry powder

• CLEANING WORK

The cleaning of any window or any transparent or translucent wall, ceiling or roof in or on a structure where such cleaning involves a risk of a person falling more than 2 metres

❑ ***Construction (Design and Management) Regulations 1994***

• CLIENT

A person or organisation who conceives and finances a construction project to be carried out under the CDM Regulations

The term client includes client's agents and developers. Further, client means any person for whom a project is carried out, whether it is carried out by another person or is carried out in-house.

❑ ***Construction (Design and Management) Regulations 1994***

• CLINICAL WASTE

Clinical waste is any waste which consists wholly or partly of human or animal tissue, blood or other body fluids, excretions, drugs or other pharmaceutical products, swabs or dressings, or syringes, needles or other sharp instruments, being waste which unless rendered safe may prove hazardous to any person coming into contact with it

It also includes any other waste arising from medical, nursing, dental, veterinary, pharmaceutical or similar practice, investigation, treatment, care, teaching or research, or the collection of blood for transfusion, being waste which may cause infection to any person coming into contact with it.

❑ ***Controlled Waste Regulations 1992***

• COCHLEA

A spiral-shaped structure which is part of the inner ear

It responds to sound vibrations and is an essential component of the hearing organism

• CODE OF PRACTICE

This is a document containing practical guidance recommended by manufactures and trade and industry associations

It does not carry as much weight as an *approved code of practice*

• CO-DECISION PROCEDURE

A legislative procedure used by the EC to achieve consensus on legislation between the European Council and the European Parliament

It is similar to the co-operation procedure in that it attempted to give the *European Parliament* more say in the determination of EU law by achieving consensus between the Council of Ministers and Parliament. Measures adopted under the procedure must be signed by the joint Presidents of both organisations and for the first time the procedure conferred a power of veto on Parliament.

❑ ***Single European Act 1986***

• COFFERDAM

A temporary dam on a construction site constructed to exclude water in order to give access to an area which is normally submerged

• COMBUSTION

Combustion is the combination of a substance with oxygen which produces a chemical reaction whereby heat and light (fire) is produced

Combustion requires a combination of fuel, heat and oxygen. ➔ *fire triangle*

• COMFORTER AND CARER

A comforter and carer is an individual who, other than part of his occupation, knowingly and willingly incurs an exposure to ionising radiation resulting from the support and comfort of another person who is undergoing or who has undergone any medical exposure

• COMMERCIAL SAMPLE

Commercial sample means, in relation to a substance or preparation dangerous for supply, a sample of that substance or preparation provided to the recipient with a view to subsequent purchase

• COMMON EMPLOYMENT

A common law expedient which allowed an employer to escape liability for injuries to one of his employees caused by the negligence of a fellow employee

It was held that workers in common employment had to accept the risk of injury arising from the negligence of co-workers. This was law until repealed by the Law Reform (Personal Injuries) Act 1948.

• COMMON LAW

One of the divisions of English law which evolved from feudal times and which was based on customs and practices which became common throughout the land – hence the term common law

• COMMON PARTS

Those parts of premises in common use

In buildings where there is multi-occupancy, areas such a lobbies, stairways etc which are used by all the occupants of the premises are said to be in common use, i.e. 'parts of premises used in common by, or for providing common services to or common facilities for the occupiers of the premises'. In an office block where there are a number of tenants, each will be responsible for the health and safety arrangements within the parts of the premises governed by his tenancy agreement. The landlord of the premises will be responsible for the health and safety arrangements within the common parts of the building.

❑ ***Health and Safety (Enforcing Authority) Regulations 1998***

• COMPARTMENT

A compartment is an enclosed space within a building in which walls, ceiling and floor consist of fire-resisting elements of building construction

• COMPARTMENTATION

Compartmentation is a method of preventing fire spread within a building by dividing the structure into separate fire-tight compartments

Each compartment will consist of fire-resisting elements of building construction in order to contain a fire within the compartment.

❑ ***BS 4422.***

• COMPENSATION

A financial settlement paid to employees who have suffered work-related injuries or diseases or in the case of fatal injuries paid to dependants

Payments can be received under social security arrangements, employers' liability insurance or in the shape of damages awarded by the courts for civil wrongs.

• COMPETENCE

An ability to perform tasks or operations to a recognised standard

Competence in health and safety terms does not necessarily depend on the possession of particular skills or qualifications. Simple situations may require only an understanding of relevant current best practice; an awareness of the limitation of one's own experience and knowledge; and the willingness and ability to supplement existing experience and knowledge. It has been defined as follows:

'A person shall be regarded as competent where he or she has sufficient training and experience or knowledge and other qualities to assist in undertaking the measures he or she needs to take to comply with the requirements and prohibitions imposed upon him or her by or under the *relevant statutory provisions.*'

❑ ***Management of Health and Safety at Work Regulations 1992***

• COMPLIANCE

The arrangements and measures necessary to satisfy legal requirements

• CONDUCTION

The transmission of heat energy through solid matter where a higher energy temperature is conducted to a lower temperature

If, for example, the end of a poker is held in a fire, the heat travels up the metal rod until the handle becomes hot. Fire spread can take place by conduction through a steel door or along a steel beam.

• CONFINED SPACE

Any place, including any chamber, tank, vat, silo, pit, trench, pipe, sewer, flue, well or other similar space in which, by virtue of its enclosed nature, there arises a reasonably foreseeable specified risk

❑ ***Confined Spaces Regulations 1997***

• CONSTRUCTION

Construction (work) is the carrying out of any building, civil engineering or engineering construction work.

It includes any of the following:

> The construction, alteration, conversion, fitting out, commissioning, renovation, repair, upkeep, redecoration or other maintenance, decommissioning, demolition or dismantling of a structure; the preparation, including site clearance, exploration investigation and excavation, and laying or installing the foundations of the structure; the assembly of prefabricated elements to form a structure or the disassembly of prefabricated elements which previously formed a structure; the removal of a structure or part of a structure or of any product or waste resulting from demolition or dismantling of a structure; or from disassembly of prefabricated elements which previously formed a structure; and the installation, commissioning, maintenance, repair or removal of mechanical, electrical, gas, compressed air, hydraulic telecommunications, computer or similar services which are normally fixed within or to a structure.

❑ ***Construction (Design and Management) Regulations 1996***

• CONSTRUCTION INDUSTRY TRAINING BOARD (CITB)

The Construction Industry Training Board is the training board for the whole of the construction industry

It provides training and certification for all aspects of industry. It can be contacted at **www.citb.org.uk.**

• CONSTRUCTION PHASE

The period of time starting when construction work in any project starts and ending when construction work in that project is completed

❑ ***Construction (Design and Management) Regulations 1994***

• CONSTRUCTIVE DISMISSAL

The termination of a contract of employment by the resignation of an employee because of circumstances caused by the employer's conduct, i.e. breach of contract on the part of the employer

• CONSULTATION

Consultation with employees is a statutory health and safety requirement placed on employers

Every employer has an obligation to consult trade union appointed safety representatives on health and safety arrangements. In non-union workplaces, the employer must consult with employees directly or through appointed representatives of employee safety on the following matters:

- introduction of any measure affecting health and safety
- the appointment of persons nominated to provide health and safety assistance, and assist in emergency procedures
- any health and safety training or information the employer is required to provide to the employees or the safety representatives
- the health and safety consequences of the planning and introduction of new technologies in the workplace
- provision of any relevant information required regarding health and safety legislation.

The employer must also inform the employees of the names of employee representatives, the group of employees they represent, and when consultation has been discontinued.

❑ ***Safety Representatives and Safety Committees Regulations 1977***
❑ ***Health and Safety (Consultation with Employees) Regulations 1996***

• CONSUMER UNIT

A combined fuseboard and main switch controlling and protecting a consumer's installation where the incoming supply of energy is delivered

• CONTAMINATION

1 The contamination by any radioactive substance of any surface including any surface of the body or clothing or any part of absorbent objects or materials or the contamination of any liquids or gases

❑ ***Ionising Radiations Regulations 1985***

2 The introduction or occurrence in food of any microbial pathogens, chemicals, foreign material, spoilage agents, taints, unwanted or diseased matter, which may compromise its safety or wholesomeness

❑ ***Food Safety (General Food Hygiene) Regulations 1995***

• CONTRACT

An agreement which is enforceable at law based on an offer by one party to another and an acceptance of the offer by that other party for a valuable consideration

➔ *invitation to treat*

• CONTRACT OF EMPLOYMENT

A legal agreement in writing between an employer and an employee

The written contract must be given to the employee within two months of the commencement of the employment..

❑ ***Employment Rights Act 1996***

• CONTRACTOR

A contractor is any person who carries on a trade, business or other undertaking (whether for profit or not) in connection with which he undertakes to or does carry out or manage construction work or arranges for any person at work under his control (including, where he is an employer, any employee of his) to carry out or manage construction work

❑ ***Construction (Design and Management) Regulations 1994***

• CONTRIBUTORY NEGLIGENCE

Where an employee has been injured at work mainly through the negligence of his employer but the employee him/herself had contributed to the accident in a minor way, e.g. by disobeying instructions, any damages awarded by the court may be reduced to the extent that the employee was partly responsible

Employees are considered to have contributed to the injury by their own actions. If the actions were the sole or major reasons for the injury then no fault or negligence could be attributed to a third party. This is covered by statute:

'Where any person suffers damage as the result partly of his own fault and partly of the fault of any other person, a claim in respect of that damage shall not be defeated by reason of the fault of the person suffering the damage, but the damage recoverable in respect thereof shall be reduced to such an extent as the court thinks just and equitable having regard to the claimant's share in the responsibility for the damage'.

❑ ***Law Reform (Contributory Negligence) Act 1945***

• CONTROL OF MAJOR ACCIDENTS

Control of major accidents involving certain amounts of hazardous substances is specifically covered by regulations

These regulations divide organisations into Lower and Top Tier Establishments depending upon the amounts of substances stored on their premises. Top Tier Establishments must prepare emergency plans for incidents both on- and off-site.

❑ ***Control of Major Accident Hazards Regulations 1999 (COMAH)***

• CONTROL LIMIT

A control limit indicates the lowest tolerable level of exposure to a hazardous substance or process which it is reasonably practicable to achieve

A control limit in relation to asbestos specifies the concentration of asbestos in the atmosphere averaged over certain time limits.

❑ ***Control of Asbestos at Work Regulations 1987***

• CONTROLLED AREA

An area in work connected with *ionising radiation*, which employers must designate as 'controlled', where dose limits are likely to be in excess of three tenths of a stipulated dose limit.

❑ ***Ionising Radiations Regulations 1985***

• CONTROLLED WASTE

Controlled waste is the normal accumulation of rubbish generated by household, commercial or industrial premises

➔ *special waste*

Waste is any scrap material, effluent, or any other unwanted surplus material arising from the application of any process and any substance or article for disposal as a result of being broken, worn, contaminated or spoiled.

❑ ***Environmental Protection Act 1990***

• CONTROL OF SUBSTANCES HAZARDOUS TO HEALTH

Control of Substances Hazardous to Health Regulations were introduced in 1992 to govern the use of substances hazardous to health in the workplace

They stipulate that employers must carry out assessments on work which is likely to expose employees to any substance hazardous to health. The regulations are one of the 'six-pack' of legislation emanating from directives issued by the EC to Member States. They were up-dated in 1999.

❑ ***Control of Substances Hazardous to Health 1999***

• CONVECTION

A process in which heat circulates in the atmosphere (gas) or in a liquid

The molecules in liquids or gases move freely. Heating a container of gas or liquid from underneath causes the molecules in the vicinity of the heat source to increase their vibrations and the fluid to expand. Expansion results in a lowering of density and the hotter portion, therefore, rises. Cold gas or liquid enters the system, receives heat and rises in turn, thus setting up the circulation of convection currents.

• COOK CHILL

Food preparation where food is prepared in advance for re-heating several days later

This requires strict control of chilled storage temperatures if food is to remain safe for consumption.

❑ ***Food Safety (General Food Hygiene) Regulations 1995***

• COOK FREEZE

Food preparation where food is prepared in advance and then deep frozen

When properly packaged such food may be kept for several months without loss of quality.

❑ ***Food Safety (General Food Hygiene) Regulations 1995***

• CO-OPERATION PROCEDURE

A European legislative procedure introduced by the Single European Act to give the European Parliament more involvement in the shaping of EU legislation

The procedure introduced two Parliamentary readings and qualified majority voting on certain issues.

❑ ***Single European Act 1986***

• CORGI

The acronym for the Council for Registered Gas Installers

The Council maintains a register of competent gas installers whose certificates of competence must be renewed every five years.

• CORPORATE LIABILITY

➔ *liability*

CORPORATE MANSLAUGHTER

The killing of an individual as a result of the reckless failure or gross carelessness of the management of an organisation

This is an issue which has not yet been resolved by legislation because of the difficulty in determining which individual director of the company or organisation can be held responsible. ➔ *liability*

• CORROSIVE

A corrosive substance, such as acids and alkalis, is one which will chemically attack materials or persons, damaging material and destroying living tissue

• COSHH

➔ *Control of Substances Hazardous to Health*

❑ ***Control of Substances Hazardous to Health Regulations 1999***

• COST BENEFIT ANALYSIS

A business technique in which the financial implications and the ensuing benefits of a project are evaluated in detail

• COTIF

The abbreviation for the Convention concerning International Carriage by Rail, as revised or re-issued from time to time

• COUNCIL OF MINISTERS

The (European) Council of Ministers is composed of one member from each of the Member States

It is the principal decision-making and legislative body of the EU.

❑ ***Article 145 EC***

• CRADLES

➔ *working platform* and *suspended access systems*

• CRANE

Lifting machinery other than a hoist or a lift

❑ ***Lifting Operations and Lifting Equipment Regulations 1998***

• CRANFIELD MAN

Research undertaken by the Cranfield Institute of Technology into the required physical attributes of the ideal operator to operate a horizontal lathe turned out to be a 'frankenstein' figure with a squat body and exceptionally long arms

This occurred because no ergonomic account was taken of the position and location of controls in relation to the physical attributes of the average operator. ➔ *ergonomics*

• CRIMINAL COURTS

Criminal Courts in the UK are those dealing with criminal offences

They are divided into two categories depending on the seriousness of the offence being tried. Summary offences may only be tried in the Magistrates'

Court; indictable offences may only be tried in the Crown Court; and some offences, known as either-way offences may be tried in either Magistrates' or Crown Court depending on the option exercised by the accused. Three magistrates sit together to hear cases. They are unpaid and not normally legally qualified but take advice from a legally qualified Clerk to the Court who sits with them. Some legally qualified magistrates – formerly called 'stipendiaries' but now known as 'district judges' – are allowed to sit alone to hear cases. They may also act as examining magistrates when they hold preliminary examinations into offences to determine whether the prosecution can show prima facie that an offence has been committed. If a prima facie case is established, the defendant can be committed for trial to the Crown Court.

Magistrates' powers are generally limited to fines of up to £5,000 and sentences of up to six months imprisonment but breaches of the general duties of the HSAW Act carry a maximum fine of £20,000 and a maximum sentence of 6 months imprisonment.

Conviction for breaches of health and safety legislation at Crown Court can carry an unlimited fine and/or imprisonment for up to two years. Crown Courts also hear appeals against conviction from Magistrates' Courts.

• CRIMINAL INJURIES COMPENSATION

A system of awarding financial compensation to victims of crime who have sustained injury as a result of the criminal offence

Awards are made by the Criminal Injuries Compensation Board based on a tariff according to the degree of injury.

• CRIMINAL LIABILITY

➔ *liability*

• CROCIDOLITE

➔ *asbestosis*

• CROWN COURT

➔ *criminal courts*

• CROSS-CONTAMINATION

The transfer of germs or bacteria from contaminated food to other foods

• CRYOGENICS

A method of refrigeration where liquefied gas is injected into a storage chamber

- **CULLEN REPORT**

A Report compiled by Lord Cullen in 1992 as a result of his enquiry into the facts surrounding the *Piper Alpha Disaster* in 1988

- **CULTURE**

The term used to describe the accepted ideas, practices, values and behaviour of a particular society or organisation

➔ *safety culture*

• DAMAGES

Monetary compensation for loss suffered by a person owing to a breach of contract or tort due to the negligence of some other person

Where a person is killed or injured at work due to the negligence of his employer there is normally an entitlement to damages. ➔ *liquidated and unliquidated damages.*

• DANGEROUS GOODS

Dangerous goods are classified as:

- explosives
- radioactive materials
- goods listed in the Approved Carriage List
- goods harmful to the environment as listed in HSWA (Application to Environmentally Hazardous Substances) Regulations 1996

• DANGEROUS OCCURRENCE

A dangerous occurrence is an occurrence arising out of or in connection with work, and includes the following:

- collapse of lifting machinery
- failure of pressure systems, e.g. boilers, pipework
- failure of freight container
- unintentional incident involving overhead electric lines exceeding 200 volts
- electrical short-circuit which stops 'plant' for more than 24 hours or is potentially life-threatening
- incidents involving explosion
- incidents involving release or potential release of biological agents likely to cause severe human infection or illness
- incidents involving malfunction of radiation generators used in radiography, irradiation of food or processing of products by irradiation

- incidents involving breathing apparatus while in use or under test in certain circumstances
- incidents in relation to diving operations which put the diver at risk
- incidents involving collapse of scaffolding more than 5 metres high or erected over or adjacent to water where there is a risk of drowning
- incidents involving train collisions
- incidents in relation to wells
- incidents in respect of a pipeline or pipeline works
- incidents involving failure of fairground equipment in use or under test
- incidents involving the carriage of dangerous substances by road
- incidents involving unintended collapse of a building or structure involving a fall of more than 5 tonnes of material
- incidents in respect of explosion or fire resulting in the stoppage of plant for more than 24 hours
- incidents involving the escape of flammable substances
- incidents involving the accidental release or escape of any substance in a quantity sufficient to cause death, major injury or any other damage to the health of any person.

❑ ***The Reporting of Injuries, Diseases and Dangerous Occurrences Regulations 1985***

• DANGER ZONE

Any zone in or around machinery in which a person is exposed to a risk to health or safety from contact with a dangerous part of machinery

• DATE OF KNOWLEDGE

In civil actions for personal injury, the plaintiff must normally commence an action for damages within three years of the injury occurring. Where ill-health arises following exposure to certain substances (e.g. asbestos) the illness may not become apparent until long after the expiry of the three year period.

As a result of this problem, a 'date of knowledge' has been introduced into the procedure. This allows the three year period to commence from the date of knowledge – the date when the victim discovers that he or she has the right to sue.

❑ ***Limitation Act 1980***

• DAY

Day means a period of 24 hours beginning at midnight

❑ ***Working Time Regulations 1998***

• DEAD

Dead, in electrical terms, means at or about earth potential and disconnected from any live system

• DECIBEL

The decibel (dB) is the unit of measurement for measuring the relative loudness of sounds

The decibel scale is not linear but logarithmic, e.g. a 20 decibel sound is ten times as loud as a 10 decibel sound and a 30 decibel sound is a hundred times as loud as a 10 decibel sound.

❑ ***Noise at Work Regulations 1989***

• DECLARATION OF CONFORMITY

A declaration by a manufacturer that machinery purchased by any EU Member State conforms to essential health and safety requirements

Any item of work equipment must be designed and constructed in compliance with any essential European Community requirements concerning the safety of products.

❑ ***Provision and Use of Work Equipment Regulations 1998***

• DECLARATION OF INCORPORATION

A declaration issued by a manufacturer that parts of machinery which are to be incorporated into other machinery comply with essential health and safety requirements

❑ ***Supply of Machinery (Safety Regulations) 1992***

• DEFECTIVE EQUIPMENT

Equipment which, for some reason, is not operating at the correct level of efficiency or safety

Defective equipment can be the cause of personal injury or loss of production in an organisation. Where an employee suffers personal injury in the course of his employment as a result of using defective equipment provided by his employer and the defect is attributable wholly or in part to the fault of a third party (whether identified or not), the injury shall be deemed to be also attributable to negligence on the part of the employer. The employee would be able to sue his employer successfully and his employer, in turn, would be able to take action against the manufacturer.

❑ ***Employer's Liability (Defective Equipment) Act 1969***

• DEFENDANT

A person against whom an appeal is brought in a civil action or the accused in a criminal prosecution

• DEFRA

The acronym for the Department for Environment, Food and Rural Affairs

The Department is responsible for policy and operational issues in relation to environmental protection, food, farming and fisheries. It came into being in June 2001 to replace the Ministry of Agriculture, Fisheries and Food (MAFF).

• DEMOLITION

Demolition is the disassembly of prefabricated elements which previously formed a structure

Demolition work is notifiable to the HSE as construction under the CDM Regulations. Suitable and sufficient steps must be taken to ensure that demolition which exposes any person to risk of danger is planned and carried out in such a way, so far as is practicable, as to prevent such danger. Demolition must be planned and carried out under the supervision of a competent person.

❑ ***Construction (Design and Management) Regulations 1994***

• DEMOUNTABLE TANK

A tank having a capacity greater than 450 litres which is not designed for the carriage of goods without breakage of load, and can normally only be handled when empty

A demountable tank is not an integral part of the vehicle and is not attached to the frame of the vehicle (whether structurally or otherwise) and, except when empty, is not intended to be removed from the vehicle.

• DEMS

The acronym for 'Disaster and Emergency Management Systems'

• DEOXYRIBONUCLEIC ACID

DNA is a substance found in the cell nucleus of all living matter which is the carrier of genetic information

DNA provides the inherited coded instructions for an organism's development and DNA analysis of body tissues and fluids is used in the identification of individuals

• DEPARTMENT OF TRADE AND INDUSTRY (DTI)

The government department for trade and industry

• DERMATITIS

An inflammation of the skin which can be caused by direct contact with a substance or agent (contact dermatitis) or by an allergic reaction (allergic dermatitis) through sensitisation to a specific substance or agent

If it occurs as a result of working conditions it is termed occupational dermatitis.

• DEROGATION

Derogation is the deviation from a set of rules or standards

For example, in the Working Time Regulations 1998, although working hours are stipulated, under certain circumstances workers do not have to comply with them.

• DESIGN

In relation to any structure, design includes drawing, design details, specification and bill of quantities (including specification of articles or substances) in relation to the structure

• DESIGNER

Any person who carries on a trade, business or other undertaking in connection with which he prepares a design or arranges for any person under his control (including, where he is an employer, any employee of his) to prepare a design, relating to a structure or part of a structure

❑ ***Construction (Design and Management) Regulations 1994***

• DETECTOR

A part of an automatic fire detection system that contains at least one sensor which constantly monitors at least one suitable physical or chemical phenomenon associated with fire

It provides at least one corresponding signal to the control and indicating equipment. ➔ *heat* and *smoke detector*

❑ ***British Standard 5839, Pt.1***

• DEVELOPER

A commercial developer who has sold domestic premises prior to completion of the project and who arranges for the construction work to be carried out

❑ ***Construction (Design and Management) Regulations 1994***

• DIOXIDE

An oxide containing two oxygen atoms

• DIOXIN

A highly-toxic organic compound produced as a by-product of certain industrial processes

• DIRECT CURRENT

An electric charge which does not alter its directional flow

• DIRECT ENTRY

One of the routes of entry of toxic substances into the body whereby entry is directly through the skin perhaps via a cut or abrasion

• DISABILITY

Disability is defined as having a physical or mental impairment affecting mobility, dexterity, co-ordination, continence or memory, causing substantial and long-term adverse effects on the ability to carry out normal day-to-day activities

'Long-term' means more than 12 months. A disabled person means a person who has a disability.

❑ ***Disability Discrimination Act 1995 (DDA)***

• DISABILITY DISCRIMINATION

Under disability discrimination legislation, employers must not discriminate against any person because of his/her disability

Discrimination occurs where an employer, for a reason which relates to the disabled person's disability, treats him or her less favourably than the employer treats or would treat others to whom that reason does not or would not apply and he cannot show that the treatment in question is justified.

The employer also discriminates against a disabled person if the employer fails to make reasonable adjustments in relation to the disabled person and cannot show that the treatment is justified. Disabled persons must be treated on equal terms with the non-disabled in relation to employment or promotion in employment opportunities. The provision of furniture and fittings in the workplace, in relation to the disabled, must also be taken into consideration by the employer.

❑ ***Disability Discrimination Act 1995***

• DISASTER PLANNING

A legal requirement under health and safety legislation

➔ *Serious and Imminent Danger* and *COMAH*

❑ ***Control of Major Accident Hazards Regulations 1999***

• DISCOVERY

The compulsory disclosure of documents by either side to each other in an action at law

• DISCRIMINATION

The act of treating persons favourably or unfavourably in comparison with others

Discrimination is an offence under a number of Acts.

❑ ***Equal Pay Act 1970***
❑ ***Sex Discrimination Act 1975***
❑ ***Race Relations Act 1976***
❑ ***Disability Discrimination Act 1995***

• DISEASE – OCCUPATIONAL

A disease which entitles an employee who is diagnosed as suffering from exposure to it, owing to working conditions, to disablement benefit

The prescribed occupational diseases are listed.

❑ ***Social Security (Industrial Diseases) (Prescribed Diseases) Regulations 1980.***

• DISINFECTION

The reduction in levels of contamination on food equipment or in food premises normally by the use of chemicals to kill micro-organisms

❑ ***Food Safety (General Food Hygiene) Regulations 1995***

● DISMISSAL

The termination of a contract of employment by an employer

➔ *constructive dismissal*

Dismissal may be fair or unfair. An employee who has been continuously employed for two years or more has a right not to be unfairly dismissed. An employee who feels he or she has been unfairly dismissed may make a complaint to that effect to an employment tribunal.

❑ ***Employment Rights Act 1996***

● DISPLAY SCREEN EQUIPMENT (DSE)

Any alphanumeric or graphic display screen, regardless of the display process involved

The definition includes the typical office VDU but also covers microfiche, liquid crystal displays, process control equipment and any other display screen used to show alphanumeric text or graphics.

❑ ***Health and Safety (Display Screen Equipment) Regulations 1992***

● DOMINO THEORY

This suggested that the events leading up to an accident are similar to a row of dominoes where, if the first one is pushed over, it knocks over the next one and so on until the whole row has collapsed

➔ *Heinrich*

Causal Events

Heinrich's Dominoes

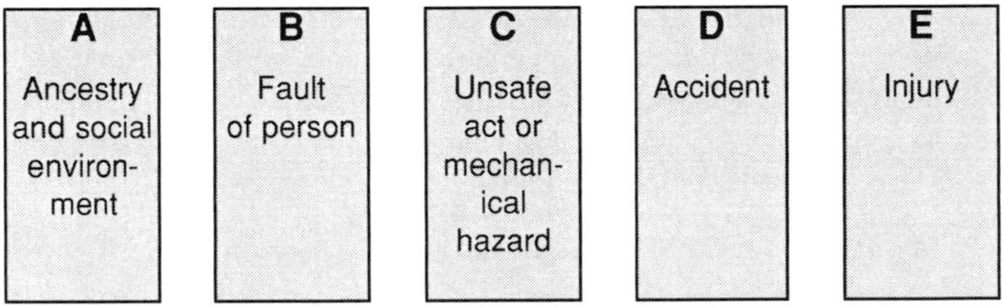

A	B	C	D	E
Ancestry and social environ-ment	Fault of person	Unsafe act or mechan-ical hazard	Accident	Injury

● 🕮 DONOGHUE v STEVENSON

The case which highlighted the common law principle that individuals must take reasonable care of others

The circumstances of the case were that a female customer consumed part of the contents of a bottle of ginger beer in a café. The bottle contained the remains of a decomposed snail. The female became ill. Her male companion had purchased the ginger beer so there was no contract between the female and the café proprietor. However, she successfully sued the manufacturer.

The judgement in the case was that reasonable care must be taken to avoid acts or omissions which, with reasonable foresight, you would know would be likely to cause injury to your 'neighbour'. This 'neighbour principle' is now widely used to establish the existence of the duty of care. The question to be asked is 'would a reasonable person in the position of the defendant have been able to contemplate or foresee harm arising from his act or omission?' If the answer is in the affirmative then a duty of care does exist.

It is now a well-established principle at common law that employers must take reasonable care to protect employees from the risk of foreseeable injury disease or death at work.

• DOPPLER EFFECT

A phenomenon associated with the frequency of sound waves

The sound from an approaching source, e.g. a police siren, shifts to a higher frequency as it approaches the listener and on passing changes to a lower frequency as it moves away. Also referred to as 'Doppler shift'.

• DOSE

A dose in radiation terms is the term used to describe the amount of radiation absorbed by a person

A minimum lethal dose is the minimum amount of a substance per unit body of an experimental animal which would cause a fatal effect. It is often expressed as LD_{50} which is the lethal dose which will kill 50% of test animals. Similarly LC_{50} is the lethal concentration of an airborne substance which will kill 50% of test animals.

• DOSE RATE

Dose rate means, in relation to a place, the rate at which a person or part of a person would receive a dose of ionising radiation from external radiation if he were at that place, being a dose rate at that place averaged over one minute

❑ ***Ionising Radiations Regulations 1999***

• DOSE RECORD

In relation to a person, the record of the doses received by that person as a result of his exposure to ionising radiation, being the

record made and maintained on behalf of the employer by the approved dosimetry service in accordance with Regulation 21

❑ ***Ionising Radiations Regulations 1999***

• DOSEMETER

An instrument used to measure the amount of radiation exposure

❑ ***Ionising Radiations Regulations 1985***

• DOSIMETER

An instrument for measuring daily exposure to noise

❑ ***Noise at Work Regulations 1989***

• DOUBLE-BARRELLED ACTION

An action in which two courses of action may arise from the same set of circumstances

For example where an employee is injured because of his or her employer's negligence in failing to guard machinery adequately, the employee may sue for damages simultaneously but separately on grounds of negligence (civil action), and breach of statutory duty (criminal action).

📖 ***Kilgollan v Cooke & Co. Ltd 1956***

• DOUBLE INSULATION

The provision of two separate layers of insulation between the live parts of an appliance and the part being handled

Double insulation appliances do not require *earthing*.

Double insulation symbol

❑ ***Electricity at Work Regulations 1989***

• DRENCHER

A fire-protection system used to protect roofs, windows and walls by discharging a curtain of water between the fire and the protected area

• DRINKING WATER

An adequate supply of wholesome drinking water shall be provided for all persons at work

❑ ***Workplace (Health, Safety and Welfare) Regulations 1992***

• DUE DILIGENCE

A defence in law available in certain circumstances to an accused who is able to demonstrate that he or she took all reasonable precautions and used due diligence to avoid the commission of the offence

• DUST

An aerosol formed by the disintegration of a solid substance

This can be by a process such as handling, sawing, grinding, drilling, detonation etc. Dust particles become temporarily suspended in the air and are dangerous to health from inhalation leading to lung disorders. They are also hazardous as concentrations of dust in the atmosphere can lead to fire and explosion.

• DUST EXPLOSION

An explosion caused by the rapid burning of flammable dust suspended in the atmosphere

• DUTY HOLDER

A person who holds a duty under legislation to carry out certain obligations

For example, under the EAW Regulations 1989, a duty holder is any person who is in control of plant or equipment which directly or indirectly can cause danger or injury to other persons.

❑ ***Electricity at Work Regulations 1989***

• DUTY OF CARE

Duty of Care is the care which a prudent and reasonable person would, with reasonable foresight, exercise towards another person

This is also referred to as the neighbour principle. It is a legal responsibility owed by one person to another not to act carelessly. If as a result of a careless act injury is caused to another person, that third person may sue for damages.

➔ 🕮 *Donoghue v Stevenson 1932*

- **EARTH**

  The conductive mass of earth whose electric potential at any point is taken as zero

- **EARTH FAULT LOOP IMPEDANCE**

  The total opposition to current flow, starting and ending at the point of fault

- **EARTH LEAKAGE CIRCUIT BREAKER (ELCB)**

  An ELCB is a device which senses the current flow to earth within an electrical circuit and disconnects the supply of electricity before a person can receive a potentially fatal electric shock

- **EARTHING**

  Connecting an electrical circuit direct to the mass of earth by means of an earth electrode in order to prevent any acquisition or build-up of charge

In an earthed circuit, the connection from appliance to earth must be continuous and must not pass through any means capable of breaking the connection.

- **EARTH TERMINAL**

  The main earthing connection point at the electrical intake position

- **EC CERTIFICATE OF ADEQUACY**

  A certificate issued by an approved body which is satisfied that an application is accompanied by a schedule containing all the required information and that vessels manufactured in accordance with the schedule would conform with a relevant national standard

• EC CERTIFICATE OF CONFORMITY

This is issued by an approved body to a manufacturer who has decided not to apply for EC verification

In such a case, the manufacturer must submit an application before the commencement of series manufacture accompanied by documentation describing the manufacturing processes and all the measures to be taken to ensure conformity with a national standard or with the relevant prototype.

• EC TYPE – EXAMINATION CERTIFICATE

A certificate issued by an approved body certifying that a prototype representative of the production envisaged satisfies the requirements of the Directive

• ECOLOGY

The relationship between living organisms and their environment

• ECO-MANAGEMENT

A system designed to promote total quality environmental management

• ECO-SYSTEM

A system in which there is an on-going relationship and interchange between living organisms and their surroundings

• EFFLUENT

Waste from residential, industrial or agricultural sources which may enter the environment

• EH40

An HSE Guidance Note published annually

The current publication is identified by the yearly suffix, e.g. EH40/03. It contains the current statutory list of substances which have been assigned either a *MEL* or an *OES*. The advice contained in EH40 must be read in conjunction with the COSHH Regulations.

• EMPLOYEE SAFETY

➔ *consultation*

• EINECS

The abbreviation for The European Inventory of Existing

Commercial Chemical Substances – a list of commercial chemical substances

❑ ***Notification of Existing Substances (Enforcement) Regulations 1994***

• ELECTRICITY

A phenomenon arising from the association of negatively and positively charged particles of matter

Atoms consist of electrons, which are negatively charged particles, and protons, which are positively charged particles. Particles with opposite charges attract each other whilst those with similar charges repel. Electrons and protons in an atom are normally in balance where the charge is equalised. Where the balance is disturbed providing a surplus of negative and positive charges, an electrical potential energy is created.

• ELECTRIC CURRENT

The measure of the rate of flow of electrons along an electric conductor

• ELECTRICAL EQUIPMENT

Anything used, intended to be used or installed for use, to generate, provide, transmit, transform, rectify, convert, conduct, distribute, control, store, measure or use electrical energy

❑ ***Electricity at Work Regulations 1989***

• ELECTRIC SHOCK

Electric shock is caused by a person coming into contact with a live part of an electrical installation

The lethal level of electric current passing through a person can be as low as 50mA.

• ELECTRICITY AT WORK

The use of electricity at work is governed by regulations covering all places of work; all electrical equipment, electrical systems; and electrical installations within the workplace

❑ ***Electricity at Work Regulations 1989***

• ELECTRICAL INSTALLATION

An assembly of associated electrical equipment

❑ ***Electricity at Work Regulations 1989***

● ELECTROLYTE

A solution with the ability to conduct an electric current

● ELECTRON

The negatively-charged particle of an atom

● ELINCS

The abbreviation for European List of Notified Chemical Substances

This is a list of new chemical substances which have been notified to the HSE as intended for commercial distribution.

❑ ***Notification of New Substances Regulations 1993***

● EMERGENCY ACTION CODE

The code required to be displayed on tanks and vehicles which are being used for the carriage of certain dangerous goods, ascertained in accordance with the *Approved Carriage List*

● EMERGENCY ESCAPE OR FIRST-AID SIGN

A sign giving information on escape routes, emergency exlts, first-aid, or rescue facilities

❑ ***Health and Safety (Safety Signs and Signals) Regulations 1996***

● EMPLOYER'S LIABILITY

The employer's duty of care towards his employees whilst they are at work engaged on his business

Every employer carrying on a business in Great Britain must insure, and maintain insurance against liability for bodily injury or disease sustained by his employees, arising out of and in the course of their employment in Great Britain in that business. The Employer's Liability Insurance Certificate must be prominently displayed on an employer's premises at each place of business where the policy holder employs persons covered by the insurance.

❑ ***Employers' Liability (Compulsory Insurance) Act 1969***

● EMPLOYMENT

Employment, in relation to a worker, means employment under his contract, and 'employed' shall be construed accordingly

❑ ***Working Time Regulations 1998***

• EMPLOYMENT APPEALS TRIBUNAL (EAT)

The Employment Appeals Tribunal operates under the jurisdiction of the Employment Tribunals Act 1996 and consists of a mixture of High Court judges and lay persons to hear appeals on decisions reached by employment tribunals

❑ ***Employment Tribunals Act 1996***

• EMPLOYMENT MEDICAL ADVISORY SERVICE

EMAS is a Service provided under the HSWA to advise and inform the Secretary of State, the Health and Safety Commission and others concerned with the health of those in employment or seeking employment, on matters relating to health in relation to employment or training for employment

❑ ***Health and Safety at Work etc Act 1974***

• EMPLOYMENT TRIBUNAL

Employment Tribunals (formerly called industrial tribunals) consist of a legally qualified chairman and two lay members who examine cases concerning employment

One lay member represents management and the other employees. These persons are usually from representative bodies, i.e. trade unions and management representative organisations. They are selected from a list kept by the Department of the Employment. Tribunals are empowered to hear cases under a number of Acts.

❑ ***Equal Pay Act***
❑ ***Trade Union and Labour Relations Act***
❑ ***Sex Discrimination Act***
❑ ***Race relations Act***
❑ ***Employment Act***
❑ ***Health and Safety at Work etc Act***

The Rules of Evidence are relaxed although evidence is given under oath or affirmation. Decisions require a simple majority vote but many decisions are unanimous. Decisions can be in favour of either party – the applicant or the respondent. The decision may:

- uphold a dismissal or order reinstatement of employment
- order payment of monies owed
- affirm or cancel improvement or prohibition notices.

Costs are not normally awarded in Tribunal cases.. The appeals procedure is through the Employment Appeals Tribunal to the Court of Appeal and to the House of Lords

❑ *Employment Tribunals Act 1996*

● ENABLING ACT

An Act of Parliament which 'enables' the making of *subordinate legislation*

● ENDOPARASITE

A parasite which lives in a host animal

● ENDOTHERMIC

A term applied to a chemical reaction in which heat is absorbed

● ENFORCEMENT

Enforcement of health and safety legislation is generally divided between the HSE for industrial premises and the local authority for commercial premises within its area

❑ *Health and Safety at Work etc Act 1974*

● ENFORCEMENT OFFICERS

Inspectors employed by the HSE or local authorities who are empowered to inspect premises and workplaces to enforce health and safety legislation

❑ *Health and Safety (Enforcing Authority) Regulations 1998*

● ENVIRONMENT

Environment consists of all, or any, of the following media, namely, the air, water and land, and the medium of air includes the air within buildings and the air within other natural or man-made structures above or below ground

❑ *Environment Act 1995*

● ENVIRONMENT AGENCY

The Environment Agency (and the Scottish Environment Protection Agency) was established as a corporate body with powers in relation to contaminated land, abandoned mines, National Parks, fisheries, the control of pollution, the conservation of natural resources, the conservation and enhancement of the environment, and the imposing of

obligations on certain persons in respect of certain products or materials.

❑ ***The Environment Act 1995***

● ENVIRONMENTAL HEALTH OFFICER

A local authority inspector whose powers are similar to an *HSE Inspector*

❑ ***Environmental Protection Act 1990***

● ENVIRONMENTAL POLLUTION

Pollution of the air, water or land which may give rise to any harm

Pollution includes pollution caused by noise, heat or vibration or any other kind of release of energy.

❑ ***Pollution Prevention Control Act 1999***

● ENZYME

A protein which acts as an organic catalyst affecting the rate of a metabolic reaction

● EPIDEMIOLOGY

The study of the frequency and distribution of an infectious process

● EPIDERMIS

The outermost layer of skin

● ERGONOMICS

The study of the man-machine interface – the relationship between the worker and the working environment

Failure to apply ergonomics in the planning and designing stages of work equipment can produce machinery which is difficult or uncomfortable to operate. ➔ *Cranfield Man*

● ESCALATOR

An escalator must be fitted with devices to prevent trapping between sides and ends of moving treads and must have readily identifiable and accessible emergency stop controls

❑ ***Workplace (Health, Safety and Welfare) Regulations 1992***

• ESCAPE ROUTE

A route forming part of the means of escape from a point in a building to a final exit

❑ *BS 4422*

• ESCHERICHIA COLI

E.coli is a bacterium of faecal origin which can lead to fatal food-poisoning

• EUROPEAN COMMISSION

The European Commission consists of Commissioners appointed by each of the Member States to formulate and implement policy decisions, to promote and represent the interests of the EU, and if necessary to institute proceedings for any violations of Community policies by Member States. Commissioners hold office for five years

❑ *Article 155 E. C.*

• EUROPEAN COURT OF HUMAN RIGHTS

Established in 1959 to deal with breaches of the European Convention on Human Rights

Judges are drawn from each of the 40 member countries. The Court sits in Strasbourg – not to be confused with the *European court of Justice*

• EUROPEAN COURT OF JUSTICE

Sits in Luxembourg and its judges are drawn from the 15 member states. Its function is to deal with breaches of European Law

• EUROPEAN ECONOMIC COMMUNITY (EEC)

The EEC was established by the Treaty of Rome with the objective of establishing a common market with no trade barriers

The abbreviation EEC was later contracted to EC – European Community – and later still to EU, the European Union.

❑ *Treaty of Rome 1957*

• EUROPEAN COMMUNITY (EC)

➔ *EEC*

● EU DECISIONS

EU Decisions may be made by the *Council of Ministers* and the *European Commission*

Decisions are binding in their entirety upon those to whom they are addressed – Member States, legal persons or individuals. Those involving financial obligations are enforceable in national courts.

● EU DIRECTIVE

These Directives may be made by the Council of Ministers and the European Commission

Directives are binding on Member States with regard to the objectives to be achieved but the method of incorporation into their respective legislation is left open. Framework Directives lay down the general objectives to be achieved and Daughter Directives specify how those goals can be achieved. Directives must then be implemented by regulations made in Member States. In the UK, this is normally by means of statutory instruments

● EU LEGISLATION

EU Legislation is Community law

Under the European Communities Act 1972, Community law was incorporated into United Kingdom law. European Community law arising from the Treaties or Regulations is therefore directly applicable in the United Kingdom without the need for any further action by the British Parliament.

The European Economic Treaty of 1957 specified five measures that may be introduced by the Council of Ministers and the European Commission: Regulations; Directives; Decisions; Recommendations and Opinions (see EU Regulations, Directives, Decisions, Recommendations and Opinions).

The so-called six-pack all started life as Directives. Directive 89/391 gave rise to the Management of Health and Safety at Work Regulations 1992. This was a Framework Directive and was accompanied by five Daughter Directives. These Daughter Directives resulted in the other five members of the six-pack and all six Directives were implemented by regulations made under Section 15 of the HSAW Act and took effect from 1 January 1993.

❑ ***European Communities Act 1972***

● EUROPEAN OMBUDSMAN

Appointed under the Treaty of Rome by the European Parliament to deal with maladministration by institutions and organisations of the European Union

EUROPEAN PARLIAMENT

The Parliamentary body representing the people of the Member States of the European Union

Elections are held every five years. Originally it had a mainly advisory role in EC affairs but has gradually assumed greater administrative power because of the co-operation and co-decision procedures.

EU RECOMMENDATIONS AND OPINIONS

May be made or issued by the Council of Ministers and the European Commission

They have no binding legal force and merely contain the views of the issuing institution.

EU REGULATIONS

May be made by the Council of Ministers and the European Commission

These apply directly in Member States. Where a conflict exists with national law, the Regulation prevails. There is no requirement to assimilate them into national law as with Directives.

EUROPEAN UNION

➔ *Treaty on European Union*

EXAMINATION 1

Examination of work equipment is an important factor in the maintenance of good health and safety and the prevention of accidents

In relation to power presses, guards and protection devices, the employer must ensure that a power press must not be put into service for the first time after installation, or after assembly at a new site or location, until it has been thoroughly examined to ensure that it has been installed correctly; it is safe to operate; and any defect has been remedied. In relation to *lifting equipment* an examination scheme must be drawn up by a competent person for such thorough examination at such intervals as may be appropriate for the purpose described in the Regulations.

- ❑ ***Provision and Use of Work Equipment Regulations 1998***
- ❑ ***Lifting Operations and Lifting Equipment Regulations 1998***

• EXAMINATION 2

Examination, in relation to a pressure system, means a careful and critical scrutiny of a pressure system or part of a pressure system, in or out of service as appropriate, using suitable techniques, including testing, where appropriate to assess its actual condition and whether, for the period up to the next examination, it will not cause danger when properly used if normal maintenance is carried out

For this purpose 'normal maintenance' means such maintenance as it is reasonable to expect the user (in the case of an installed system) or owner (in the case of a mobile system) to ensure is carried out independently of any advice from the competent person making the examination.

❑ ***Carriage of Dangerous Goods (Classification, Packaging and Labelling) and Use of Transportable Pressure Receptacles Regulations 1996***

• EXCAVATION

Excavation includes any earthwork, trench, well, shaft, tunnel or underground working

Excavations must be adequately protected by suitable shoring in relation to depth and width, the nature of the surrounding ground and the strains imposed by the sub-soil. Any excavated soil or building materials should be stored away from the edges of any excavation. The excavation should be lit at night and where depth exceeds 2m, suitable guard-rails should be erected.

Before commencement of any excavation tests should be made to determine the position of any underground services.

❑ ***Construction (Design and Management) Regulations 1994***

• EXOTHERMIC

A chemical reaction in which heat is emitted

• EXPLOSIVE

An explosive is a substance which has a violent, explosive effect

In order to acquire and keep a specified amount of explosives, an employer must obtain from the police a valid explosives certificate certifying that he or she is a fit and proper person to be allowed to acquire and keep explosives. The certificate must specify the amount of explosives, the purpose for which they are intended, and the specified place for storage.

❑ ***Control of Explosives Regulations 1991***

• EXPLOSIVE SUBSTANCE

A solid or liquid substance; or a mixture of solid or liquid substances or both, which is capable by chemical reaction in itself of producing gas at such a temperature and pressure and at such a speed as could cause damage to surroundings or which is designed to produce an effect by heat, light, sound, gas or smoke or a combination of these as a result of non-detonative self-sustaining exothermic chemical reactions

• EXPLOSION SUPPRESSION

A technique used to prevent an explosion by means of an automatic suppression system fitted to plant or machinery

The system detects the advent of the explosion and immediately injects the suppressant (inert gas, powder, vaporising liquid, etc.) into the plant which extinguishes the explosion flame.

• EXPLOSION VENT

An opening in a vessel which is covered by a flap or disc (see *bursting disc*) which in the event of an explosion in the vessel allows the pressure or gas to escape from the system.

• EXPOSURE LIMITS

➔ *MEL* and *OES*

• EYESIGHT TEST

Where a person is a user under the Display Equipment Regulations, or an employee who is to become a user, his or her employer must ensure that he or she is provided, on request, with an appropriate eyesight test and/or eye test to be carried out by a competent person

Employees are entitled, but have no obligation, to undergo an eye and eyesight test. Where the user opts only for an eyesight test, use of an eye-screening instrument is adequate for that purpose. However, this type of eyesight testing is not a substitute for a full eye test of visual capability. The test should assess visual capability at the normal viewing distance for a display screen. Where the result indicates that vision requires correction, the user should be advised to consult an ophthalmic optician for a full examination.

❑ ***Health and Safety (Display Screen Equipment) Regulations 1992***

• EYE TEST

Where a user requests an eye test, rather than an eyesight test, the employer must refer him or her directly to an ophthalmic optician or a medical practitioner

The employee should be tested by a competent person of the employer's choice and the cost of any test for work with a display screen must be at the employer's expense. The results of any sight test may only be disclosed with the consent of the person tested. If, during the course of an eye examination, a disease or injury is suspected, the user must be referred to a registered medical practitioner.

❑ ***Health and Safety (Display Screen Equipment) Regulations 1992***

• FACILITIES FOR CHANGING CLOTHING

Sufficient and suitable facilities shall be provided for any person at work to change clothing in all cases where the person has to wear special clothing for the purpose of work and the person cannot, for reasons of health or propriety be expected to change in another room

These facilities shall not be suitable unless they include separate facilities for use by men and women where necessary for reasons of propriety.

❑ *Workplace (Health, Safety and Welfare) Regulations 1992*

• FAILURE MODE AND EFFECTS ANALYSIS

FMEA is a management technique for predicting the potential consequences of failure in a particular component of a system or structure

It is useful in the designing of a planned maintenance system

• FALLS FROM HEIGHT

Falls from height which are likely to cause personal injury must be prevented by providing suitable platforms, ladders and scaffolding

Where the fall involves a drop of more than 2m suitable guard-rails and toe-boards must be provided.

❑ *Workplace (Health, Safety and Welfare) Regulations 1992*
❑ *Construction (Health, Safety and Welfare) Regulations 1996*

• FARADAY CAGE

An open-sided cube which is covered in a conducting material and insulated from the floor

When the cube is charged with an electric current, a person can move freely inside it and touch the sides without any adverse effect.

• FAULT

Fault means negligence, breach of statutory duty or other act or omission which gives rise to liability in tort in England and Wales or which is wrongful and gives rise to liability in damages in Scotland

❑ ***Employer's Liability (Defective Equipment) Act 1969***

• FAULT TREE ANALYSIS

A technique which examines an incident or accident starting from the direct cause and examining in turn all the preceding underlying causes and contributing factors

• FENCING

A barrier surrounding dangerous machinery which prevents persons from reaching the dangerous parts of the machine

❑ ***Provision and Use of Work Equipment 1999***

• FIBRILLATION

Spontaneous, rapid contractions of individual muscle fibres

Fibrillation of the heart muscle causes the heart to beat irregularly

• FINAL EXIT

The last point in an escape route beyond which persons are no longer in danger from fire

❑ ***BS 4422***

• FINE

A financial punishment imposed by a court on an offender who has been convicted of an offence

For breaches of health and safety legislation dealt with at Magistrates' Court, the normal maximum fine is £5,000. For certain offences, the maximum can be £20,000. For breaches heard at Crown Court there is no maximum limit on fines

• FIRE CERTIFICATE

A certificate issued by a local Fire Authority in respect of certain factories, offices and shops

The certificate certifies that the premises have been inspected by the Fire

Authority and have been rated in terms of fire risk and that the use of the premises, the means of escape, and the fire precautions in situ are adequate for the safety of the persons using the premises. Employers must apply for a fire certificate where:

- the premises are in multi-occupation
- more than 20 persons are employed at any one time
- more than 10 persons are employed elsewhere than on the ground floor
- explosive or highly flammable materials are stored or used in the premises
- the premises are hotels and boarding houses where sleeping accommodation is provided for guests or staff for 6 or more persons or at basement or above first floor level.

❑ ***Fire Precautions Act 1971***

• FIRE CERTIFICATE – SPECIAL PREMISES

In addition to fire certificates required under the 1971 Fire Precautions Act, there are a number of special premises which require fire certificates

These are premises where the manufacture of certain substances and liquids are carried out or storage of certain substances and liquids in excess of prescribed limits takes place.

These include highly flammable liquids, expanded cellular plastics, liquefied petroleum gas, liquefied natural gas, methyl acetylene, liquid oxygen, chlorine, artificial fertilisers, ammonia, phosgene, ethylene oxide, carbon disulphide, acrylonitrile, hydrogen cyanide, ethylene and propylene.

Special premises also includes explosives factories or magazines which are required to be licensed under the Explosives Act 1875 and certain other specified buildings

❑ ***Fire Certificates (Special Premises) Regulations 1976***

• FIRE CHECK DOOR

A door designed to prevent the spread of smoke or hot gases for a period of at least 20 minutes

• FIRE DOOR

A door provided for the passage of persons which together with its frame and furniture is intended when closed to resist the passage of fire or gaseous products of combustion

Its function is to protect escape routes from the effects of fire and to limit the spread of fire throughout a building. A fire door and its frame should have a fire resistance of at least 20 minutes and be fitted with non-combustible hinges with a high melting point. Greater fire-resistance can be achieved by using intumescent strips.

❑ ***BS 4422***

• FIRE EXTINGUISHER

Hand-held, first-aid, fire-fighting equipment containing a variety of agents according to the classification of fire

• FIRE LOAD

The total amount of combustible material on premises expressed in heat units

❑ ***BS 4422***

• FIREMAN'S SWITCH

An emergency switch fixed in a conspicuous position in certain buildings and premises, usually adjacent to a main entrance or exit, which is capable of isolating all live conductors in the premises

The switch is coloured red and sited in a position agreed with the fire authority

• FIRE-RESISTING WALL

A load-bearing or non load-bearing wall which is capable of satisfying for a given period of time the criteria of fire resistance with regard to collapse, flame penetration and excessive temperature rise

• FIRE RISK ASSESSMENT

Employers are required to assess the risk of fire occurring in the workplace and the risk to people in the event of a fire occurring in the workplace

A fire risk assessment will embody the following factors:

- identification of potential fire hazards
- identification of persons at risk in the event of fire
- evaluation of the risks
- recording of the findings and action taken
- informing employees of the results of the assessment

- reviewing and revising assessment as and when necessary.

❑ ***Fire Precautions (Workplace) Regulations 1999***

• FIRE SAFETY SIGN

A sign, including an illuminated sign or an acoustic signal, which provides information on escape routes and emergency exits in case of fire; provides information on the identification or location of fire-fighting equipment; or gives warning in case of fire

❑ ***Health and Safety (Safety Signs and Signals) Regulations 1996***

• FIRE SPREAD

The growth and spread of fire across the surface of a structure

• FIRE STOP

A physical barrier designed to restrict the spread of fire in cavities within a building

The space around the holes cut in walls between adjoining rooms to enable central heating pipes or electrical conduits to pass through should be fire stopped to prevent fire spreading through the aperture.

• FIRE TRIANGLE

A simple representation of the combustion process where the three sides of the triangle are represented by oxygen, heat and fuel

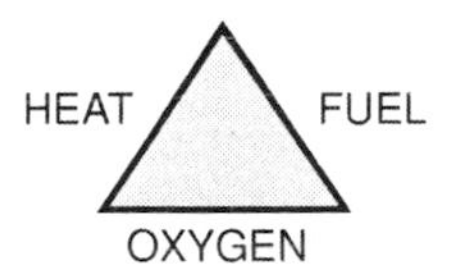

If any one of the three sides is absent, then combustion is not possible.

• FIRE VENT

A ventilator in the walls or roof of a building which may be operated automatically or manually in the event of fire in order to release heat or smoke from the building

❑ ***BS 4422***

• FIRST AIDER

A person qualified to provide basic medical treatment prior to the arrival of professional medical assistance

Employers must have facilities for first-aid in their places of work. In addition to facilities, there should be qualified first-aiders to provide basic medical treatment prior to the arrival of professional medical assistance. The number of first-aiders required is assessed in relation to a number of factors, e.g. the number of employees; the degree of occupational risk involved; the company's accident record; and the distance from emergency medical services. ➔ *appointed persons* and *suitable person*

❑ ***Health and Safety (First-Aid at Work) Regulations 1981***

• FLAMEPROOF EQUIPMENT

The term applied to apparatus which is enclosed in a case which can contain any spark or flame within it without igniting any flammable gas present in the surrounding atmosphere

• FLAMMABLE

Referring to the capacity of a substance to catch fire easily

• FLAMMABLE LIQUIDS

Highly flammable liquids are any liquids, such as paraffin, petrol, white spirit etc., with a flash point below 32° C

HFL must be stored in fixed storage tanks in safe positions or in suitable closed vessels in a safe position in the open air, protected from sunlight. If kept in a storeroom, unless it is impracticable to do so, every storeroom, cupboard, bin, and tank used for storing HFL must be clearly marked 'highly flammable' or 'flashpoint in range of 22° to 30°', or with some other clear indication of the degree of flammability. If kept in a workroom, the HFL must not exceed 50 litres and must be stored in suitable closed vessels and kept in a fire-resisting structure.

❑ ***Highly Flammable Liquids and Liquefied Petroleum Gases Regulations 1972***

• FLAMMABILITY

The capacity to burn with a flame

• FLASH POINT

The lowest temperature at which a liquid produces flammable vapour in air in a sufficient quantity to produce a flash when a small flame is applied

The lower the flash point temperature, the more readily ignition will occur.

• FLOCCULATION

A process by which small particles form into larger clumps or masses

• FLOORS

Floors must be suitable for their intended purpose, e.g. storage, pedestrians, traffic etc.

Floors must be properly constructed, not overloaded, and kept free from obstruction.

❑ ***Workplace (Health, Safety and Welfare) Regulations 1992***

• FLUID

Fluid means gases, liquids and vapours in pure phase as well as mixtures thereof; a fluid may contain a suspension of solids

❑ ***Pressure Equipment Regulations 1999***

• FLUID IN GROUP 1

A dangerous fluid – a substance or preparation covered by the definitions of Article 2 of Council Directive 671548/EEC relating to the classification, packaging and labelling of dangerous substances as specified as:

- explosive
- extremely flammable
- highly flammable
- flammable (where the maximum allowable temperature is above flashpoint)
- very toxic
- toxic, or
- oxidising.

❑ ***Pressure Equipment Regulations 1999***

• FLUOROCARBON

A synthetic organic compound in which some or all of the hydrogen atoms have been replaced by fluorine atoms

• FOOD LEGISLATION

The Food Safety Act covers such matters as unfit food, unhygienic premises and the training of workers and persons handling food

It provides for the inspection of food premises by authorised inspecting officers who have the power to issue *improvement and prohibition orders*. The Food Safety Hygiene Regulations apply more specifically to the training, instruction and supervision of food handlers.

- ❑ ***Food Safety Act 1990***
- ❑ ***Food Safety (General Food Hygiene) Regulations 1995***

• FORKLIFT TRUCK (FLT)

A truck used for the lifting and transporting of materials.

The load is carried on lifting forks projecting forward of the front wheels and counterbalanced by weight located behind the rear wheels. (see *reach truck*). Drivers must be over 18 years of age and trained under an approved scheme

- ❑ ***Supply of Machinery (Safety Regulations) 1992***

• FORMALDEHYDE

A pungent, colourless gas which is toxic by contact, inhalation and ingestion and is carcinogenic

It is used in a number of manufacturing processes.

• FOSSIL FUELS

Non-renewable sources of energy such as coal, oil and natural gas

Their use involves the release of carbon dioxide into the atmosphere which can lead to air pollution and other environmental problems.

• FRAGILE MATERIAL

Any material which would be liable to fail if the weight of any person likely to pass across or work on that material (including the weight of anything for the time being supported or carried by that person) were to be applied to it

- ❑ ***Construction (Health, Safety and Welfare) Regulations 1996***

• FRAGILE ROOF

Roofs which are of flimsy construction which are unlikely to support the weight of materials or workers

Where work is to take place on such roofs, warning notices must be exhibited and access must only be by way of suitable scaffolding.

- ❑ ***Construction (Health, Safety and Welfare) Regulations 1996***

• FRANGIBLE DISC

A disc, usually metal, incorporated in a safety channel which is designed to burst and relieve excess pressure

• FREE RADICAL

An atom or group of atoms containing an unpaired *electron*

Free radicals have a very short-lived existence and, according to some researchers, contain some chemicals produced as a by-product of certain biological activity which are harmful to healthy cells. This is thought to affect the ageing process and is possibly a contributing factor in the development of cancer.

• FREQUENCY RATE

➔ *accident data*

• FUME

Fume is created by the process of conversion by heating of a solid to a molten state

Metallic fumes can be highly toxic.

• FUMIGATION

To fumigate is to disinfect an area by exposing it to fumes

It is defined as an operation in which a substance is released into the atmosphere so as to form a gas to control or kill pests or other undesirable organisms.

❑ ***Control of Substances Hazardous to Health Regulations 1999***

• FUSE

An overload and short-circuit protection device

A fuse consists of a short length of wire housed in a small cartridge. The fuse is placed between the supply and the cables connected to the load or equipment and the fuse wire is rated at the same current. If the current is exceeded, the wire will melt thereby interrupting the supply of electricity.

• FUSIBLE LINK

A soldered connecting link designed to melt at a specific temperature

It can be used, for example, as a fire-door release device which holds the door-closing mechanism in the open position. When the link breaks in a heat situation, the mechanism then operates to close the door

• GAMMA

Gamma (γ) rays, in radiation, are electromagnetic waves which can travel long distances and are capable of passing through the body and causing severe damage to cell tissue

➔ *ionization*

❑ ***Ionising Radiations Regulations 1985***

• GAS

One of three phases of any solid, i.e. solid, liquid or gas

A gas is any material which is in a gaseous state at normal pressure and temperature.

• GAS CONTAINERS

Gas containers (transportable) are defined as containers (plus their permanent fittings) of between 0.5 litres and 3,000 litres whose main purpose is for transporting the contents

❑ ***Pressure Systems and Transportable Gas Containers Regulations 1989***

• GATES

Gates must be of suitable construction including being fitted with any necessary safety device

Any sliding gate must have a device to prevent it coming off its track and any upward opening gate must have a device to prevent it falling back. Any powered gate must have suitable effective features to prevent it causing injury by trapping any person.

❑ ***Workplace (Health, Safety and Welfare) Regulations 1992***

• GEIGER-MULLER TUBE

A measuring device used to measure, by the process of ionisation, the amount of radiation present in the atmosphere

• GENERIC ASSESSMENT

➔ *risk assessment*

• GENETIC MODIFICATION

The alteration of genetic material otherwise than by mating or natural recombination or both

• GLAZING

The complete or partial covering of an aperture with transparent or translucent material

Glazing in the workplace must be of safe and sound construction.

Every window or transparent or translucent surface in a door or gate must be of safety material or protected against breakage and be appropriately marked. Any window, skylight or ventilator which is capable of being opened must not be likely to be opened, closed or adjusted in a way which exposes the person opening it to a risk to health and safety. Windows, skylights and ventilators when open must be positioned in such a way that no person in the workplace is exposed to a risk to health and safety. All windows and skylights in a workplace must be of a design or construction that allows them to be cleaned safely. ➔ *safety glass*

❑ ***Workplace (Health, Safety and Welfare) Regulations 1992***

• GLOBAL WARMING

A theory that the Earth's climate is becoming warmer

➔ *greenhouse effect*

• GRAB SAMPLING

An atmospheric monitoring technique to determine the amount (if any) of any harmful airborne contaminants in the air to which employees may be exposed

The technique draws a quantity of air through a glass tube containing a substance which becomes discoloured if the air contains certain impure substances. ➔ *stain tube*

• GREENHOUSE EFFECT

A theory attributed to the increase in so-called greenhouse gases – carbon dioxide, methane, nitrous oxide and chlorofluoro-carbons – as a result of the burning of fossil fuels, deforestation, etc.

The 'greenhouse' theory suggests that these gases, which form a blanket in the atmosphere surrounding the Earth, are increasing to an extent which reduces the amount of the Sun's energy, which is normally reflected into space. Some researchers believe that this increased 'blanketing' is contributing to a warming of the global climate.

• GREEN PAPER

A consultative document issued at the preliminary stage in proposing legislation

The proposals are published in a green-coloured document and circulated for discussion amongst MPs and other interested parties. ➔ *white paper*

• GUARDS

Guards and protection devices for machinery must be suitable for the purpose for which they are provided and be of good construction, sound material and adequate strength

Guards must be maintained in an efficient state, in efficient working order and in good repair. They must not give rise to any increased risk to health or safety and not be easily by-passed or disabled. They must be situated at a sufficient distance from the danger zone and not unduly restrict the view of the operating cycle of the machinery, where such a view is necessary.

They must also be so constructed or adapted to allow replacement operations and maintenance work to take place restricting access except to the area where the work is to be carried out and, if possible, without having to dismantle the guard or protection device.

❑ ***Provision and Use of Work Equipment Regulations 1999***

• GUARDING

Guarding machinery is subject to a hierarchy of measures under Regulations which recommend the following order of precedence:

- the provision of fixed guards
- the provision of other guards (automatic, interlock, trip) or protection devices

- the provision of jigs, holders, push sticks or similar protection appliances
- the provision of information, instruction training and supervision.

❑ ***Provlsion and Use of Work Equipment Regulations 1999***

• GUARD RAIL

Horizontal cross-members fitted to a scaffold or working platform

Guard rails must be fitted where it is possible to fall from a height of 2 metres or more. The top guard rail must be fitted 1100mm above the scaffold boards (walkway) and the intermediate rail half-way between the top rail and the toe board. ➔ *scaffold.*

• GUIDANCE BOOKLETS

Publications issued by the HSE to provide health and safety advice and guidance.

They have no legal standing and are not so authoritative as an approved code of practice. They cover issues such as general chemical safety; plant and machinery; and medical and environmental hygiene.

➔ *Appendix 4* for some of the current publications

• HACCP

The abbreviation for *Hazard Analysis Critical Control Point* – a management factor used in *HAZOPS* for the identification and control of hazards

• HALF-LIFE PERIOD

The time in which the radioactivity of a substance drops to one half of its original value

It is important in determining the time scale in which pollutants decay and are expelled from a biological system

• HALON

A hydrocarbon

Halon gas was formerly used in computer suites as a fire-extinguishing agent. As halon contains CFC, now banned by the *Montreal Protocol*, it is being replaced by a halon substitute, or in some cases, water.

• HANDLING

Manual handling operations means any transporting or supporting of a load (including the lifting, putting down, pushing, pulling, carrying or moving thereof) by hand or by bodily force

❑ ***Manual Handling Operations 1992***

• HAND SIGNAL

A movement or position of the arms or hands or a combination thereof, in coded form, for guiding persons who are carrying out manoeuvres which create a risk to the health and safety of persons at work

➔ b*anksman*

• HAND VIBRATION

Hand arm vibrating syndrome is caused by prolonged use of vibrating hand tools

The continued exposure to vibration causes constriction of the blood vessels in the hands. ➔ *vibration white finger.*

• HARMONISED STANDARD

A technical specification adopted by the European Committee for Standardisation or the European Committee for Electrotechnical Standardisation or both, and of which the reference number is published in the Official Journal of the European Communities

❑ ***Pressure Equipment Regulations 1999***

• HAWTHORNE EXPERIMENT

A series of management researches into human relations in the workplace, conducted at the Western Electric Company (USA) in the 1920s

The research attempted to evaluate the effect of better lighting in the workplace on production by placing a number of operatives in a better working environment than those in the main factory. The findings noted an improvement in production but were flawed because the group taking part were aware that they were under observation; members of the group bonded together; and the researchers and the group under observation developed a rapport which did not exist between the normal workforce and management.

Although the lighting experiment did not (at that time) prove a relationship between better illumination and production, the flawed experiment led to a realisation that informal workgroups and friendly supervision were significant factors in improving employee output.

• HAZARD

Anything – machinery, tools, substance, system of work etc. or person – which has the potential to cause harm

• HAZARD ANALYSIS CRITICAL CONTROL POINT

A management tool giving a structured approach to the identification and control of hazards

It involves the employment of a multi-disciplinary expert team.

• HAZCHEM

The HAZCHEM Code is a hazardous chemicals warning system

Chemicals being transported in tanks and containers can be identified by a warning board displayed on the vehicle which contains coded information about the hazardous substance such as an emergency action code, a substance identification number, and an emergency telephone number for assistance. ➔ *TREMCARD, Appendix 5*

❑ ***Carriage of Dangerous Goods by Road Regulations 1996***

• HAZOPS

The acronym for 'hazard and operability study', a technique involved in the assessment of potential hazards which might arise, for example, from the individual components of a new chemical installation or engineering plant

The study is managed by a team of experts in safety, engineering, chemistry, production etc depending on the nature of the process under examination. The technique looks, at the design stage, at the potential outcome of incorrect installation, operation or malfunctioning of these individual items of equipment and how such malfunction could affect the facility as a whole.

• HAZARD WARNING PANEL

Road vehicles which are transporting hazardous substances must carry a hazard warning panel to indicate the nature of the hazardous substance and the action to be taken in the event of an emergency such as a road accident

The panel is orange in colour and displays the emergency action code; the substance identification number; a specialist advice telephone number; and the appropriate hazard warning sign or symbol for the substance.
➔ *Appendix 5*

• HEALTH AND SAFETY CASE LAW

A number of decided cases have had a considerable influence on the interpretation of health and safety law

The following cases are in approximate date order and not in order of importance.

Rylands v Fletcher 1868

Strict liability – liability for damage caused by escape of 'dangerous things' from land or property

Smith v Baker 1891

Volenti non fit injuria – voluntary acceptance of risk. In personal injury circumstances, difference between knowledge of risk and consent to run risk.

Donoghue v Stevenson 1932

Negligence – duty of care – neighbour principle. The duty of care exists if it is reasonably foreseeable that a person's acts or omissions are likely to cause harm to another person

Wilsons & Clyde Coal Company Ltd v English 1938

Employer's duty to employees to include the provision of competent staff, adequate material, a proper system and effective supervision

Mersey Docks and Harbour Board v Coggins and Another 1947

Vicarious liability – accident caused by first company's crane operator where both crane and operator had been hired by another company

Edwards v National Coal Board 1949

Explanation of 'reasonably practicable' – balancing quantum of risk against the degree of sacrifice involved in avoiding the risk.

Paris v Stepney Borough Council 1951

Negligence – employer's duty owed to each individual workman

Latimer v AEC Ltd 1953

Negligence and 'reasonable practicability' in relation to unforeseen or freak hazards in the workplace

Adsett v K and L Steelfounders and Engineers Ltd 1953

'Practicable' in relation to state of knowledge at the time, available resources and knowledge of scientific experts

Marshall v Gotham Co Ltd 1954

Further interpretation of the meaning of 'reasonably practicable'

Summers and Sons Ltd v Frost 1955 -

Absolute duty to fence dangerous parts of machinery

Richard Thomas and Baldwins v Cummings 1955

Machinery – guidance on 'in motion and in use '

🕮 Kilgollan v Cooke & Co Ltd 1956

'Double-barrelled action' where a claim for negligence and breach of duty arise out of same set of circumstances, i.e. claim for damages (civil action) and prosecution for breach of duty (criminal offence)

🕮 Lister v Romford Ice and Cold Storage Co Ltd 1957

Employee's breach of contractual duty to employer to take due care and exercise skill

🕮 Davie v New Merton Board Mills Ltd 1959

Injury to employee caused by faulty equipment supplied by employer not the fault of the employer.

🕮 McWilliams v Sir William Arrol & Co Ltd 1962

Provision of safe system of work and liability in relation to link between defendant's conduct and plaintiff's injury

🕮 Rookes v Barnard 1964

Award of exemplary damages

🕮 Uddin v Associated Portland Cement Mfrs Ltd 1965

Machinery guarding – guidance on 'safe by position'

🕮 J H Dewhurst Ltd v Coventry Corporation 1969

Machinery – cleaning of dangerous machinery by young persons

🕮 British Railways Board v Herrington 1972

Occupier's liability in relation to trespassers – in particular, duty of common humanity towards child trespassers

🕮 Rose v Plenty 1976

Vicarious liability – an employer's liability – employee doing something unauthorised in the course of his employment

🕮 Armour v Skeen 1977 -

Safe system of work and liability of a body corporate, or senior persons in an organisation, for offences under the HSAWA 1974

🕮 White v Pressed Steel Fisher Ltd 1980

Guidance on attendance at union-organised health and safety training courses

🕮 Regina v Swan Hunter Shipbuilders and Telemeter Installations Ltd 1981

Responsibilities of all parties and arrangements for their management when contractors and sub-contractors are working together on site and employers' duty to inform and instruct others, as well as employees, of potential dangers

🕮 Thompson v Smith Ships Repairers Ltd 1984

Safe system of work – acceptance of risk – occupational hearing loss – date of knowledge

🕮 Williams and West Wales Plant Hire Co and Others 1984

Guidance on law relating to hire of equipment and hire of equipment operator

🕮 Walker v Northumberland County Council 1994

Duty to provide a reasonable safe system of work included the duty not to cause the risk of psychiatric damage

🕮 R v Associated Octel Ltd 1994

Interpretation of Section 3 of HSWA 1974 and the meaning of the phrase ' conduct of an undertaking'

• HEALTH AND SAFETY FILE

A record of information for the *client*, which focuses on health and safety

It alerts those who are responsible for the structure and equipment in it to the significant health and safety risks which will have to be dealt with during subsequent use, construction, maintenance, repair and cleaning work.

❑ ***Construction (Design and Management) Regulations 1994***

• HEALTH AND SAFETY LEGISLATION

For principal legislation ➔ Appendix 1

• HEALTH AND SAFETY PLAN

Documentation, in construction work, which serves two purposes

At the pre-tender stage, a health and safety plan, prepared before the tendering process, brings together the health and safety information obtained from the client and designers and aids in the selection of the *principal contractor.*

During the construction phase, the health and safety plan details how the construction work will continue to be managed to ensure health and safety.

❑ ***Construction (Design and Management Regulations) 1994***

● HEALTH SURVEILLANCE

A requirement under the MHSW Regulations to ensure that employees are provided with such health surveillance as is necessary having regard to any risks to their health and safety identified in any risk assessment

A number of other specific regulations, e.g. COSHH, Lead at Work, Asbestos, place on employers, whose employees are exposed to certain occupational health risks, a requirement to carry out health surveillance. For example, where employees are exposed to certain levels of asbestos, they must undergo a medical examination at least every two years. Such medical examination records must be kept for at least 40 years owing to the lengthy incubation period of the disease. Medical surveillance has been defined as including clinical assessment and biological monitoring.

❑ ***Management of Health and Safety at Work Regulations 1999***

● HEARING

➔ *noise* and *presbycusis*

● HEAT DETECTOR

Devices designed to detect fire in its more advanced stage when the temperature in the affected area starts to increase

They are often sited at ceiling height in premises and operate in two ways, i.e. a fixed temperature detector which operates when a pre-determined temperature is reached and a rate of rise detector which operates when the rise in temperature is more rapid than normal. In both cases the operation sets off an alarm. ➔ *smoke detector*

● HEAT SINK

A device in a system where unwanted heat can be stored or dissipated.

Used extensively in electronic equipment such as computers.

● HEINRICH

H W Heinrich was an accident causation theorist

➔ *domino effect*

In his book, *Industrial Accident Prevention*, he posited the theory that there was a relationship between accidents and unsafe acts. He estimated that in a total group of 330 accidents of the same kind, there would be: 1 major or lost time injury; 29 minor injuries; 300 no injury accidents.

• HEPATITIS

An inflammation of the liver

Hepatitis A is commonly caused by consuming water or food contaminated by faeces. Hepatitis B is transmitted sexually or by inoculation of contaminated blood, e.g. needle-sharing amongst drug abusers. Viral hepatitis is a prescribed disease reportable under *RIDDOR*.

• HMSO

Her Majesty's Stationery Office

The printers of all UK legislation including health and safety statutes and regulations. HMSO can be contacted at **www.hmso.co.uk.**

• HSC

The abbreviation for the Health and Safety Commission

The HSC was established by the HSWA 1974 and consists of a chairman and 6 to 9 other members appointed by the Secretary of State. Its functions in relation to health and safety matters are, amongst others, to encourage and carry out research; provide training and information; to provide an information and advisory service to government departments, employers' and employees' representative organisations; to direct investigations and inquiries to submit proposals to the Secretary of State for the making or revocation of regulations; and to ensure that directions given to it from the Secretary of State are effectively carried out.

• HSE

The abbreviation for the Health and Safety Executive

The HSE was established by the HSWA 1974 and consists of three persons appointed by the HSC with the approval of the Secretary of State. Its main function is to carry out and give effect to any of the HSC directions and to make adequate arrangements for the enforcement of health and safety regulations

• HSE INSPECTOR

These inspectors are appointed by HSE or local authorities under the HSWA

The appointment is made in writing specifying their powers. They may only exercise the powers specified and only within the field of responsibility of the authority which appointed them. Inspectors must produce, on demand, their instrument of appointment.

An inspector may:

- enter any premises in order to exercise his or her powers
- be accompanied by a police officer if there is reasonable cause to believe serious obstruction might be met in the execution of duty
- make any necessary examinations and investigations
- direct that those premises or part of them or anything in them, e.g. machinery, shall be left undisturbed for the purpose of investigation or examination
- take photographs and such measurements as are deemed necessary
- take samples, e.g. dust, of any articles or substances found in any premises
- dismantle any article or substance which is likely to cause danger to health and safety or subject them to any process or test
- retain any article or substance to ensure it is available for use as evidence in any court proceedings
- require any person, whom he or she has reasonable grounds for believing may be able to give relevant information about the investigation, to answer questions and to sign a declaration as to truth of his or her answers
- require the production of, inspect, and take copies of, or of any entry in, any books or documents which are required to be kept by regulations
- having taken possession of any article or substance, leave a notice at the premises with a responsible person stating that he or she has taken possession: if that is not practicable, the notice must be fixed in a conspicuous place; where a sample has been taken, the inspector shall give a portion of it to a responsible person
- issue improvement or prohibition notices.

❑ ***Health and Safety at Work etc Act 1974***

• HERTZ

Hertz (Hz) is a unit of sound frequency indicating the number of cycles per second

• HIERARCHY OF NEEDS

➔ *Maslow*

● HIGHLY FLAMMABLE LIQUID

Highly flammable liquids (HFL) are any liquids, liquid solutions, etc., with a *flash point* below 32° C

All HFL must be stored in fixed storage tanks or in suitable closed vessels in safe positions. If in the open air, it must be protected against sunlight. If kept in a workroom, the HFL must not exceed 50 litres and must be stored in suitable closed vessels and kept in a fire-resisting structure.

❑ ***Highly Flammable Liquids and Liquid Petroleum Gases Regulations 1972***

● HIV

The human immunodeficiency virus which may lead to AIDS

● HOIST

A platform or cage whose direction or movement is limited or restricted by a guide or guides

❑ ***Factories Act 1961***
❑ ***Lifting Operations and Lifting Equipment Regulations 1998***

● HOMEOSTASIS

The state within an organism that allows it to maintain a stable internal environment

● HOMEWORKER

A person who is employed to work at home by an employer

Such a person is deemed to be 'at work' from a health and safety point of view.

● HOUSEHOLD WASTE

Household waste other than:

(a) asbestos
(b) waste from a laboratory
(c) waste from a hospital, other than waste from a self-contained part of a hospital which is used wholly for the purposes of living accommodation.

❑ ***Special Waste Regulations 1996***

● HOUSEKEEPING

A term used to cover activities involving cleanliness of the workplace, furniture, fittings, floors, walls, ceilings, mopping-up of spillages, and disposal of waste and rubbish

• HUMAN FACTORS

A combination of attitudes, motivation training, human error, and perceptual, mental and physical limitations which affect the interaction between an individual and his or her job

Individual employees at all levels in any organisation have different habits, skills, personalities, knowledge, physical and mental capabilities all of which exert an influence on safety.

• HUMIDITY

Atmospheric humidity is the amount of water vapour in the air

Together with temperature and ventilation, it is a factor in determining the condition and quality of the air. In relation to DSE Regulations, the relative humidity at workstations should be in the range 40-60%

❑ ***Health and Safety (Display Screen Equipment) Regulations 1992***

• HYDROMETER

An instrument used to measure the specific gravity of liquids

• HYGROMETER

An instrument used to measure the humidity of the air

IEE

The abbreviation for the Institute of Electrical Engineers

IEE WIRING REGULATIONS

Recommendations issued by the Institute of Electrical Engineers regarding the safe selection and construction of wiring installations

❑ ***IEE Wiring Regulations 16th Edition***

IGNITION

The process which initiates combustion

Sources of ignition in the workplace are smoking, and discarded smoking materials, poor housekeeping, electrical appliances and plugs, static electricity, chemicals, radiant heat, and dust.

IGNITION TEMPERATURE

The lowest temperature at which a substance will ignite

ILLUMINATED SIGN

A sign produced by a device made of transparent or translucent materials which are illuminated from the inside or the rear in such a way as to give the appearance of a luminous surface

❑ ***Health and Safety (Safety Signs and Signals) Regulations 1996***

IMMUNE RESPONSE

The body's innate response to attack from invading micro-organisms by producing antibodies or *immunoglobulins*

IMMUNE SYSTEM

The body's natural defences against infection

• IMMUNOGLOBULIN

A class of blood protein which functions as an antibody

• IMPEDANCE

The total resistance to the flow of alternating current measured in *ohms*

• IMPROVEMENT NOTICE

A formal notice served by a HSE Inspector on an employer requiring him or her to remedy a specific contravention of health and safety procedure.

➔ *prohibition notice*

❑ ***Health and Safety at Work etc Act 1974***

• INCHING

A machine-maintenance operation whereby dangerous machinery which normally must be guarded at all times is permitted to be operated without guards in place for the purpose of cleaning and adjusting

The control button allows the machinery to move 75mm at a time followed by a pause. It then has to be actuated again to permit further movement.

• INCIDENT

According to the HSE, an incident includes all undesired circumstances and near misses which have the potential to cause accidents

➔ *accident*

• INCIDENT RATE

➔ *accident data*

• INDEPENDENT SCAFFOLD

One which has two rows of standards – vertical supporting scaffold tubes or poles

• INDUSTRIAL TRIBUNAL

➔ *Employment Tribunal*

• INDUSTRY ADVISORY COMMITTEE

The Industry Advisory Committee advises the HSC on industry standards.

• INFESTATION

The invasion and colonisation of premises by insects or vermin

This is especially important in premises where food is stored, prepared or sold, and precautions must be taken to prevent ingress by unwanted flying or walking creatures.

• INFLAMMATORY RESPONSE

One of the body's responses to physical injury or infection

➔ *immune response*

Inflammation is redness and swelling at the injury site accompanied by a release of white blood cells which assist in repairing any damage tissue.

• INFORMATION

Information is an important factor in maintaining good health and safety in the workplace

The Health and Safety at Work etc Act places a requirement upon employers to provide adequate information to employees. This has been repeated more explicitly in modern EU- based legislation.

❑ ***Management of Health and Safety at Work Regulations 1999***

• INFRA-RED

Infra-red radiation is part of the spectrum of electro-magnetic radiation

It is a non-ionizing radiation and invisible to the naked eye. Exposure to infra-red can cause burns and cataracts.

• INGESTION

The process of swallowing

It is one of the routes of entry by which toxic substances may gain access to the human body through the gastro-intestinal tract.

• INHALATION

One of the routes of entry by which airborne toxic substances may gain entry to the human body, through the process of breathing

• INJURY REPORTING

Major injuries sustained during the course of employment must be reported to the appropriate enforcing authority as soon as possible followed by a written report within ten days

❑ ***Reporting of Injuries, Diseases and Dangerous Occurrences Regulations 1995***

• INORGANIC WASTE

Waste which is not of biological origin including alkalis, cyanides, sulphides, heavy metals and most acids

• INSPECTION 1

An inspection, in relation to pressure receptacles, is a visual or rigorous inspection by a competent person as is appropriate for the purpose

The purpose is to identify whether equipment can be operated, adjusted and maintained safely and that any deterioration, defect, or damage can be detected and remedied.

❑ ***Carriage of Dangerous Goods (Classification, Packaging and Labelling) and Use of Transportable Pressure Receptacles Regulations 1996***

• INSPECTION 2

Inspection, in relation to the safety of work equipment, means that an employer must ensure that, where safety depends on the installation conditions, the equipment is inspected after installation and before being put into service for the first time, or after assembly at a new site or in a new location, to ensure that it has been installed correctly and is safe to operate

Where the equipment is exposed to conditions causing deterioration liable to result in dangerous situations, it must be inspected to ensure that health and safety conditions are maintained and that any deterioration can be detected and remedied in good time.

The inspections must take place:

- at suitable intervals
- any time that exceptional circumstances might jeopardise the safety of the work equipment.

An employer must ensure that inspection results are recorded and kept until the next inspection is due. He must also ensure that no work equipment leaves his undertaking, or if obtained from elsewhere, is used in his undertaking, unless accompanied by physical evidence that the necessary inspection under this regulation has been carried out.

❑ ***Provision and Use of Work Equipment Regulations 1998***

• INSPECTORS

➔ *HSE Inspectors*

• INSULATION

An electrical term used to describe the protection afforded to a conductor by surrounding it with non-conducting material

The rubber or pvc covering of an electric cable insulates it from contact with other equipment or persons thereby minimising the risk of shock or short-circuiting. ➔ *double insulation*

• INSURANCE

➔ *Employer's liability insurance*

• INTERNATIONAL SAFETY RATING SYSTEM

A system for evaluating the standards and levels of health and safety in the workplace

• INTRINSIC SAFETY

A term applied to any *circuit* where any sparking as a result of normal working is incapable of causing an explosion in a flammable atmosphere

Intrinsically safe apparatus is equipment constructed in such a way as to prevent explosion of any vapour caused by sparking

• INTUMESCENT STRIP

A narrow lining in a door edge or door jamb which is coated with intumescent paint which expands when exposed to high temperature thereby sealing the gap between door and jamb to prevent passage of heat, flame and smoke

➔ *fire door*

• INVITATION TO TREAT

An invitation to make an offer

It is very often simply a request for information and is not in itself an offer and does not form a basis for contractual relations. For example, a supermarket which displays goods marked at a certain price is inviting the public to make an offer. The price tag is merely an indication of the price likely to be accepted. It is not binding on the supermarket to sell at that price, or at all.

→ *contract*

• ION

An atom or group of atoms which has become electrically charged by gaining or losing electrons

• IONISATION

The process by which neutral atoms are converted to electrically charged atoms

• IONISING RADIATION

The transfer of energy in the form of particles or electromagnetic waves of a wavelength of 100 nanometres or less or a frequency of 3×10^{15} hertz or more capable of producing ions directly or indirectly

It is emitted by radioactive substances and can penetrate the human body causing tissue damage. → *alpha, beta* and *gamma* rays, *x-rays* and *non-ionizing radiation*

External radiation means, in relation to a person, ionising radiation coming from outside the body of that person; internal radiation means, in relation to a person, ionising radiation coming from inside the body of that person.

❑ ***Ionising Radiations Regulations 1999***

• IOSH

The abbreviation for the Institute of Occupational Safety and Health

IOSH is one of the main institutions offering occupational health and safety qualifications – the IOSH Diploma being one of the more sought after.

• IPC

The abbreviation for integrated pollution control which is administered and enforced by the Environment Agency

It is concerned with industrial processes whose emissions might have a pollutant effect on any environmental sector – air, land or water. → *APC*

❑ ***Environmental Protection Act 1990***

• IRRITANT

A substance or liquid which when it comes in contact with the skin causes an inflammatory reaction such as dermatitis or when inhaled (such as dust) causes irritation or longer-lasting damage to the lungs, e.g. fibrosis

• ISO

The abbreviation for the International Standards Organisation

• ISOLATION

Isolation, in electrical terms, is the making safe of a circuit by cutting it off from the source of electrical energy and locking-off the main switch

➔ *lock-off*

• IUPAC

The International Union of Pure and Applied Chemistry

• JOB SAFETY ANALYSIS

A method of analysing a task and breaking it down into an operating sequence

It is then possible to examine, at each stage, any required accident prevention measures which are considered necessary and the behavioural factors which might exert an influence on the way in which the task is performed. That behaviour could exert either an adverse or a beneficial influence. Its main value is in the design of safe systems of work but generates, as by-products, safety instructions and safety training.

• JOINT TORTFEASOR

Joint tortfeasors are persons who have committed the same wrongful act

They are jointly and individually responsible for the whole damage caused and redress can be obtained from all or each of those involved. ➔ *tortfeasor*

• JOULE

A unit of energy

In electrical terms, a joule is equal to one watt-second – the amount of energy released in one second by a current of one ampere through a resistance of one ohm.

• JUDICIARY

A collective term for judges or the system of courts responsible for the administration of justice

In the UK, there are two distinct court systems – those dealing with criminal and those dealing with civil matters.

Criminal – the magistrates' court is the lowest of the criminal courts. Magistrates hear cases and impose sentences for summary offences. The Crown Court deals with the more serious crimes known as indictable offences. The court is presided over by a High Court Judge, a Circuit Judge or a Recorder and trial takes place in the presence of a jury. The Criminal Division of the Court of Appeal hears appeals from the lower courts against sentences and convictions. The House of Lords hears appeals from decisions of the Court of Appeal (Criminal Division).

Civil – The County Court is the lowest of the civil courts dealing with claims for damages under £50,000. The High Court (Civil Division) deals with claims in excess of £50,000.

• JUDICIAL PRECEDENT

Part of the common law where, once a decision of a higher court has been given the status of a precedent, it must be followed by a lower court where the same point of law is involved

A precedent is binding on a lower court. It is this use of judicial precedent which gives rise to case law.

- Decisions of the European Court of Justice on matters of EU law are binding on all courts and tribunals in the UK.
- Precedents of the House of Lords are binding on all inferior courts, both criminal and civil.
- Precedents of the Court of Appeal (Criminal Division) are binding on the High Court and Crown and Magistrates Courts.
- Precedents of the Court of Appeal (Civil Division) are binding on the High Court, County Court, Employment Appeal Tribunal and Employment Tribunal.
- Precedents are binding on all inferior courts – High Court precedents on Crown and Magistrates, and Crown precedents on Magistrates. Magistrates' Courts decisions have no binding precedence.

On the civil side, High Court precedents bind County Courts and Employment Tribunals. Decisions of Employment Tribunals do not create precedents but are 'persuasive'.

➔ *binding precedent*

• KELVIN

A measure of temperature – one of the base units of the SI system

A Kelvin degree is equal to a Celsius degree with the scale adjusted so that zero (273°K) represents absolute zero (0°C)

• KINETIC ENERGY

The energy of motion or potential energy

• KINETIC HANDLING

A phrase used to describe the kinetic principles applied in the manual handling of loads – correct posture, knowledge of levers, and assessment of tasks, loads, environment and physical capability

• LABELLING

➔ *CHIP*

• LADDERS

Ladders used in construction work must be well-maintained and secured at top, sides and bottom

They should have level and firm footings and extend at least 1m above roof/landing place. Extending ladders should overlap by at least 3 rungs. Ladders should be placed at the correct angle to the wall, e.g. 1 foot 'out', 4 feet 'up'. There should be a landing point for rest purposes every 9m (30').

❑ ***Construction (Health, Safety and Welfare) Regulations 1996***

• LASER

The acronym 'laser' comes from Light Amplification by Stimulated Emission of Radiation. A laser is a concentrated band of wavelengths travelling in the same direction which form a highly- focused light beam

Lasers are used extensively in, amongst other things, surgery, computer technology, and fire detection.

• LAW

Law is a rule of human conduct imposed upon and enforced among the members of a given state in order to ensure some kind of social order

UK law is divided into common law and statute law. Common law evolved from feudal times based on customs and practices which became common throughout the land, hence the term common law

Statute law is written law produced through the parliamentary process. Statutes are Parliamentary legislation which supersede all other forms of law. Only Parliament can make, modify revoke or amend statutes.

• LEDGER

The horizontal tubing forming part of a *scaffold.*

• LEGIONNAIRES' DISEASE

A pneumonia-type infection caused by bacteria which accumulates in water service systems which are heated within a certain temperature range (20-45°C)

The infection can prove fatal and precautions must be taken, including the keeping of records, to control the potential for the build-up of the bacteria by regular testing of storage tanks and associated pipe-work.

- ❑ ***Control of Substances Hazardous to Health Regulations 1999***
- ❑ ***Notification of Cooling Towers and Evaporative Condensers Regulations 1992***

• LEGIONELLA

Legionella pneumophila is the bacteria which causes Legionnaires' Disease

• LEGIONELLOSIS

➔ *Legionnaires' Disease*

• LEPTOSPIROSIS

➔ *Weil's Disease*

• LETHAL CONCENTRATION (LC_{50})

The concentration of a substance which, if inhaled, would be expected to kill 50% of a population of experimental animals during an exposure time of four hours

• LETHAL DOSE (LD_{50})

The quantity of a substance that would be fatal to 50% of test animals exposed by routes other than inhalation – death occurring within 30 days of exposure

• LIABILITY

Being subject to a legal obligation which can be criminal or civil according to whether it is enforced in criminal or civil court

Criminal liability is incurred when a breach of the criminal law takes place. The person responsible for the breach is liable and must suffer the consequences. If the text of the law is written in absolute terms where duties and obligations are qualified by 'shall' or 'must' the offender has little room to manoeuvre, eg.,

'Every employer shall ensure that work equipment is so constructed or adapted as to be suitable for the purpose for which it is to be used or provided'.

Where an employer has 'failed to ensure etc', as required, he is patently in breach of the law and therefore liable.

Civil liability arises from an act or omission (a *tort*) which gives one individual or body the right to present a legal claim against another for restitution. A tort in English law is a breach of a duty imposed by law. It applies in a number of areas – nuisance and trespass being particularly common – but for the purposes of health and safety, the most important is the tort of negligence.

Absolute liability is often referred to as strict liability and applies equally in criminal and civil law. It arises when an offence governed by absolute liability is committed. In such cases, whether the person committing had good reason to commit or did not intend to commit the offence is immaterial.

Corporate liability – a corporation has a separate legal personality and can as a result be guilty of a criminal offence. The problem in prosecution lies in the fact that to prove the guilt of a criminal offence, the prosecution must prove that the offender had a 'guilty mind or guilty intent'. It is difficult in the case of a board of directors to identify which individuals are in fact the board's mind and will

Vicarious liability. 'Vicarious' means 'in place of another'. In employment law, an employer is liable for the wrongful acts or omissions of an employee provided the acts or omissions occur within the scope of employment, i.e. at work or arising from work. In such cases, the claimant (injured party) can sue both the employee personally and the employer who is vicariously liable. In practical terms, the employer is usually sued as he is more likely to be in a position to pay damages than an employee as, in any case, he is covered by insurance..

Employers' liability is covered by the Employers' Liability Insurance Acts. Amongst other provisions, the sum insured has been raised from £2m to £5m and inspectors are empowered to require production of past certificates as well as the current one. Certificates are now required to be kept for 40 years.

• LIFT

An appliance serving specific levels, having a car moving along rigid guides or along a fixed course, and inclined at an angle of more than 15 degrees to the horizontal and intended for the transport of persons or persons and goods

❑ ***Lifts Regulations 1997***

• LIFTING EQUIPMENT

Work equipment for lifting or lowering loads including its attachments used for anchoring, fixing or supporting it

❑ ***Lifting Operations and Lifting Equipment Regulations 1998 (LOLER)***

• LIFTING TACKLE

Work equipment for attaching loads to machinery for lifting, including chains, hooks, rings, shackles, swivels and ropes

❑ ***Lifting Operations and Lifting Equipment Regulations 1998***

• LIGHTING

Good lighting is an important factor in minimising risks to health and safety in a variety of workplaces

Different lighting levels, measured in lux, are required for different working environments and work processes.

❑ ***Electricity at Work Regulations 1989***
❑ ***Guidance Note HSG(G)38 – Lighting at Work***

• LIQUIDATED DAMAGES

Liquidated damages occur in breach of contract cases where the amount of damages payable for the breach have been pre-determined

The stated sum is recoverable by the injured party. ➔ *damages* and *unliquidated damages*

• LIQUID PETROLEUM GAS

LPG is commercial butane, commercial propane and any mixture of these substances

❑ ***Highly Flammable Liquids and Liquid Petroleum Gases Regulations 1972***

• LIVE WORKING CRITERIA

The precautions specified by Regulation for any work that has to be carried out on live electrical installations

Live working may only be carried out where it is not practicable to carry out the work with the installation dead. It may only be done by competent staff who have been given the necessary information about the system and who have been supplied with insulated tools and equipment. They must wear or use any required protective clothing or equipment and if possible be accompanied by another competent person.

❑ ***Electricity at Work Regulations 1989***
❑ ***HSG(G)38 – Lighting at Work***

• LOAD

'Load' includes any person and any animal

❑ ***Manual Handling Operations Regulations 1992***

• LOBBY

A protected lobby is a lobby having an adequate degree of fire protection and which forms part of the horizontal escape route

❑ ***BS 4422***

• LOCAL EXHAUST VENTILATION

LEV is a ventilation system which is sited at the point where dust or other particulates is emitted from machinery or by a process

Air is introduced into the vicinity and the particulates are collected from the operator's breathing zone and transported via a ducting system into a receptacle.

❑ ***HSG54: The Maintenance, Examination and Testing of Local Exhaust Ventilation***

• LOCKING-OFF

A safety procedure observed by electricians working on electrical circuits where the supply switch is some distance from the point of the actual work being carried out

As the switch is not under the electrician's control it is locked in the off position until the work is completed. ➔ *isolation*

❑ ***Electricity at Work Regulations 1989***

• LONE WORKER

A lone worker as the name implies is a worker who for the largest part of his/her working day works unaccompanied.

The term covers a large number of occupations and is important from the employers' point of view in that the duty of care and the need for risk assessments are not reduced and must not be neglected.

• LONG-TERM EXPOSURE LIMIT

→ *maximum exposure limit*

• LUMINAIRE

A lighting fitting that includes all the necessary parts for supporting, fixing and protecting lamps, together with any necessary control gear

• LUX

The unit of measurement used to express the intensity of light

• MACHINE GUARDING

A requirement where any machinery is dangerous and can cause injury when in operation

An employer must ensure that effective measures are taken to prevent access to any dangerous part of machinery or to stop the movement of any dangerous part of the machinery before a person enters a danger zone. The measures required are:

- the provision of fixed guards enclosing every dangerous part, but where this is not practicable
- the provision of other guards or protection devices, but where this is not practicable
- the provision of jigs, holders, push-sticks or similar protection appliances used in conjunction with the machinery but where this is not practicable
- the provision of information, instruction, training and supervision.

All guards and protection devices provided must:

- be suitable for the purpose for which they are provided
- be of good construction, sound material and adequate strength
- be maintained in an efficient state, in efficient working order and in good repair
- not give rise to any increased risk to health or safety
- not be easily bypassed or disabled
- be situated at sufficient distance from the danger zone
- not unduly restrict the view of the operating cycle of the machinery, where such a view is necessary
- be so constructed or adapted that they allow operations necessary to fit or replace parts and for maintenance work, restricting access so that it is allowed only to the area where the work is to be carried out and, if possible, without having to dismantle the guard or protection device.

MACHINERY

Machinery designed for use or operation by persons at work or machinery designed for use otherwise than at work in non-domestic premises made available to persons at a place where they may use the machinery so provided for their use there

❑ *Supply of Machinery (Safety) Regulations 1992*

MAGISTRATE

A magistrate (also called a Justice of the Peace) is normally a lay (legally unqualified) person appointed to the local magistrates' bench to deal with summary criminal offences

Cases are heard by two or more Magistrates sitting together who are advised on legal matters by the Clerk of the Court who is legally qualified. Magistrates are unpaid. A stipendiary magistrate (district judge) is a qualified solicitor or barrister who may hear cases alone. Stipendiary magistrates are salaried.

MAGISTRATES' COURT

➔ *criminal courts*

MAINTENANCE

An important factor in the prevention of accidents in the workplace

In relation to machinery, an employer must ensure that work equipment is maintained in an efficient state, in efficient working order, in good repair, and where any machinery has a maintenance log, the log is kept up to date.

❑ *Provision and Use of Work Equipment Regulations 1998*

MAJOR ACCIDENT

An occurrence (including in particular, a major emission, fire or explosion) resulting from uncontrolled developments in the course of the operation of any establishment and leading to serious danger to human health or the environment, immediate or delayed, inside or outside the establishment, and involving one or more dangerous substances

❑ *Control of Major Accident Hazards Regulations 1999*

MAJOR ACCIDENT PREVENTION POLICY (MAPP)

A document setting out the policy in respect of the prevention of major accidents which is designed to guarantee a high level of protection for persons and the environment by appropriate

means, structures and management systems

❑ ***Control of Major Accident Hazards Regulations 1999***

• MAJOR INJURIES

Major Injuries are defined as follows:

- any fracture, other than to the fingers, thumbs or toes
- amputations
- dislocation of the shoulder, hip, knee or spine
- loss of sight (whether temporary or permanent)
- a chemical or hot metal burn to the eye or any penetrating injury to the eye
- any injury resulting from an electric shock or electrical burn (including any electrical burn caused by arcing or arcing products)) leading to unconsciousness or requiring resuscitation or admittance to hospital for more than 24 hours
- any other injury leading to hypothermia, heat-induced illness or to unconsciousness, requiring resuscitation, or requiring admittance to hospital for more than 24 hours
- loss of consciousness caused by asphyxia or by exposure to a harmful substance or biological agent
- either of the following conditions which result from the absorption of any substance by inhalation, ingestion or through the skin – acute illness requiring medical treatment, or loss of consciousness
- acute illness which requires medical treatment where there is reason to believe that this resulted from exposure to a biological agent or its toxins or infected material.

❑ ***Reporting of Injuries, Diseases and Dangerous Occurrences Regulations 1995***

• MANDATORY SIGN

A sign prescribing behaviour, i.e. the information on the sign must be obeyed

❑ ***Health and Safety (Safety Signs and Signals) Regulations 1996***

• MANUAL CALL POINT

A device for the manual operation of an electrical fire alarm system

❑ ***BS4422***

• MANUAL HANDLING

Manual handling operations means any transporting or supporting of a load (including the lifting, putting down, pushing, pulling, carrying or moving thereof) by hand or by bodily force

❑ ***Manual Handling Operations Regulations 1992***

• MASLOW

A H Maslow was an American psychologist who suggested that we all, by our very nature, seek security and stability

An uncertain environment generates feelings of insecurity. In his 'Hierarchy of Needs', Maslow suggested that security forms one of the foundations of the structure. If it is not fully satisfied the rest of the structure is affected.

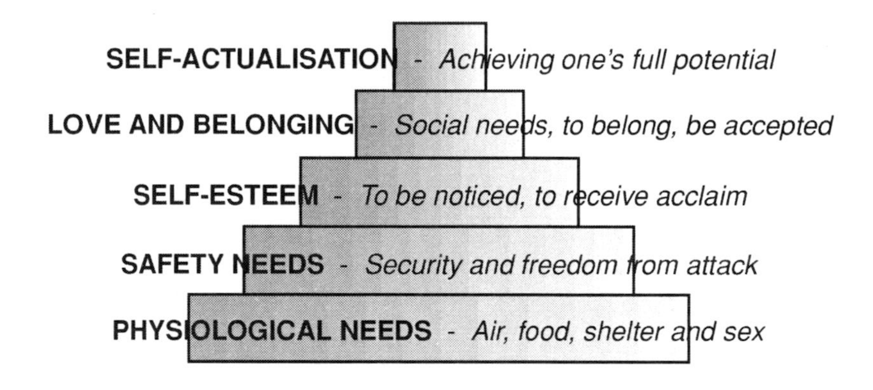

Maslow's 'Hierarchy of Needs'

Self-actualisation was the goal – the highest need – but unimportant unless the lower, more basic needs have been satisfied first.

• MATERNITY

Employers must take special account of the needs of women employees of childbearing age or new or expectant mothers who may be at risk because of a specific work process, working conditions, or a physical, chemical or a biological agent

New and expectant mothers are defined as employees who are pregnant, who have given birth or miscarried within the previous six months, or who are breastfeeding. The employer must, by a specific risk assessment, identify the specific risk posed to the health and safety of pregnant women and new mothers in the workplace.

The main risks are physical, biological, and chemical agents, and working conditions. Having identified the risk and brought them to the attention of the employee, the employer must remove the hazard or reduce the risk to its minimum. If there is a residual risk, the employer must temporarily adjust a pregnant employee's working conditions or hours, or offer suitable alternative work. If neither of these alternatives is acceptable, the employer must suspend the employee on full pay for as long as is necessary to protect her health and safety or that of her child. These 'maternity' provisions apply to all employees regardless of length of service with an employer.

❑ ***Management of Health and Safety at Work (Amendment) Regulations 1994***

• MAXIMUM ALLOWABLE PRESSURE (PS)

The maximum pressure for which the equipment is designed, as specified by the manufacturer, defined at a location specified by the manufacturer, being the location of connection of protective or limiting devices or the top of equipment or, if either of the foregoing is not appropriate, any point specified by the manufacturer

❑ ***Pressure Equipment Regulations 1999***

• MAXIMUM (or minimum) ALLOWABLE TEMPERATURE (TS)

The maximum or minimum temperatures, as the case may be, for which the equipment is designed, as specified by the manufacturer

❑ ***Pressure Equipment Regulations 1999***

• MAXIMUM EXPOSURE LIMIT

MEL is the maximum concentration of an airborne substance to which an employee may be exposed by inhalation, averaged over a reference period

The reference periods are 8 hour time-weighted (long-term exposure limits) or 15 minute time-weighted (short-term exposure limits). Employers have a duty to reduce exposure to below the MEL as far as is reasonably practicable.
➔ *OES.*

• MAXIMUM PERMISSIBLE DOSE

The maximum amount of ionising radiation to be absorbed by an individual over a given time period

❑ ***Ionising Radiations Regulations 1985***

• MEANS OF ESCAPE

Means of escape from a building are the structural means whereby a safe route is provided for persons to travel from any point in a building to a place of safety by their own unaided efforts

❑ ***BS 4422***

• MEDICAL EXPOSURE

Exposure of a person to ionising radiation for the purpose of his medical or dental examination or treatment which is conducted under the direction of a suitably qualified person and includes any such examination for legal purposes and any such examination or treatment conducted for the purposes of research

❑ ***Ionising Radiations Regulations 1985***

• MELANOMA

Skin cancer caused by over-exposure to ultraviolet radiation

• MENS REA

The Latin phrase for 'guilty mind'

It is the state of mind which illustrates that an offender in the commission of a criminal offence had the intention to carry out the act. ➔ *actus rea*

• MESOTHELIOMA

A tumour of the pleura (the lining of the lungs) brought about by exposure to dusts such as asbestosis

Mesothelioma is usually fatal. It is a prescribed occupational disease.

❑ ***Reporting of Injuries, Diseases and Dangerous Occurrences Regulations 1995***

• METHOD STATEMENT

A written description of the steps to be taken in the carrying out of a particular work activity including the health and safety measures to be taken

Specifically, a method statement is a document prepared by a *contractor* under the CDM Regulations outlining how each phase of the work he is responsible for is to be carried out. When the CDM project is completed, the method statement will form part of the *health and safety file*.

❑ ***Construction (Design and Management) Regulations 1994***

• MEZZANINE

A part-floor in between two other floors of a building

• MICRO-ORGANISM

A microbiological entity, cellular or non-cellular, which is capable of replication or of transferring genetic material

Yeast, fungi, viruses, bacteria, algae, etc are examples.

• MICROWAVE

A short electromagnetic wave which occurs in the radiation spectrum between radio wave and infra-red

• MIST

Mist consists of finely suspended droplets usually formed by the condensation of a gas or the atomising of a liquid

➔ *aerosol*

• MOBILE ELEVATING WORK PLATFORM (MEWP)

A work platform capable of being moved from one location to another

Its main use is in short duration tasks where a ladder would be unsuitable and the erection of a scaffolding platform impractical

• MONTREAL PROTOCOL

An international agreement made in 1987 which banned the production of CFC by the year 2000

➔ *ozone layer*

• MOTHERS – NURSING

➔ *maternity*

• MULTI-OCCUPANCY

The phrase 'multi-occupancy 'occurs in the MHSW Regulations in relation to the obligation placed on employers whose workers

are employed in buildings with a number of other organisations whose operations might expose the employees to risks to health and safety

The employers must inform their employees of those risks.

❑ ***Management of Health and Safety at Work Regulations 1999***

• MULTIPLEX

A method whereby several signals can be transmitted simultaneously along the same wire

• MUTAGEN

A substance with the capacity to effect genetic change or mutation in an organism

• MUTAGENISIS

A change in the genetic material in a body cell which may cause cancer or a hereditary problem – it can be brought about by exposure to ionising radiation

- **NARCOSIS**

  An anaesthetic effect brought about by exposure to a toxic substance or drug

- **NATIONAL CHEMICAL EMERGENCY CENTRE**

  The NCEC is an organisation which provides chemical emergency response products and services to health and safety professionals

- **NEAR MISS**

  An incident that does not result in accident or injury

- **NEBOSH**

  The abbreviation for the National Examination Board in Occupational Safety and Health

- **NEGLIGENCE**

  Neglect to do something

In health and safety terms, it is the neglect of the **duty of care**. Neglect can be by an act or an omission.

Negligence can be defined as the failure to take action which in those circumstances the reasonable person would take, or omitting to do something that a reasonable person would do, under the circumstances, i.e. a failure to take reasonable care. This failure to take reasonable care is founded in the common law concept of the duty of care. Each of us owes a duty of care towards others. Neglect of this duty of care is negligence. If negligence results in injury or damage then a claim for compensation could be initiated.

In order to successfully sue for negligence, three main points must be established:

- that the defendant owed a duty of care to the claimant (injured person)
- that the duty of care had been breached
- that as a result of the breach the claimant suffered injury or damage.

• NEIGHBOUR TEST

➔ 🕮 *Donoghue v Stevenson*

• NEUTRON

A sub-atomic particle containing no electrical charge

• NEWTON

A unit of force which gives to a mass of one kilogram an acceleration of one metre per second per second

• NICEIC

The abbreviation for the National Inspection Council for Electrical Installation Contracting

The Council is an independent regulatory body established to protect consumers from incompetent and unsafe electrical installations.

• NIGHT TIME

Night time, in relation to a worker, means a period which is not less than seven hours, and which includes the period between midnight and 5am, which is determined for the purposes of these Regulations by a relevant agreement, or, in default of such a determination, the period between 11pm and 6am

❑ ***Working Time Regulations 1998***

• NIGHT WORKER

A worker who works at least three hours of his daily working time during the night time, or who is likely, during night time, to work at least a proportion of his annual working time as may be specified in a collective agreement or a workforce agreement

Night work, generally speaking, is work carried out by employees between midnight and 5a.m.

❑ ***Working Time Regulations 1998***

• NIMBY SYNDROME

NIMBY is the acronym for the phrase 'not in my backyard'

The syndrome occurs where a person wants the benefits of a certain facility but does not want the facility itself sited close to that person's neighbourhood, e.g. wishing to benefit from the use of mobile telephones but not wanting mobile telephone masts sited adjacent to one's home or place of business

• NO FAULT LIABILITY

No fault liability refers to the payment of compensation to workers injured as a result of their employment where there is 'no fault' attributed to the employer

Industrial injury benefit paid under the social security system is based on a no fault liability system.

• NOISE

Unwanted sound which enters the ear by sound pressure pulses

Constant exposure to noise above certain levels can permanently damage hearing. (see *noise action levels*)

• NOISE ACTION LEVELS

Levels of noise in the workplace which require employers to take action to prevent or minimise the risk of injury to hearing of workers exposed to the noise

Employers must make a noise assessment to determine the level of noise in the workplace. There are three action levels:

- 1st noise level of 85 dB(A)
- 2nd noise level of 90 dB(A)
- 3rd noise level (peak action level) of 200 dB(A)

❑ ***Noise at Work Regulations 1989***

• NOISE LEVELS

Noise levels can only properly be determined by means of a noise survey carried out by a competent person using the correct noise measurement instruments and techniques

An indication of sound intensity is indicated by the table below

Sound Source	Sound Level	Pain Threshold
Jet aircraft	140 dB (A)	Harmful Range
Riveting hammer	130 dB (A)	
Propellor aircraft	120 dB (A)	Critical Range
Heavy motor vehicle	90 dB (A)	
Private car	70 dB (A)	Safe Range
Conversation	60 dB (A)	
Whisper	30 dB (A)	
Threshold of hearing	0 dB (A)	

• NOMINAL SIZE or DN

A numerical designation of size which is common to all components in a piping system other than components indicated by outside diameters or by thread size

It is a convenient round number for reference purposes which is only loosely related to manufacturing dimensions and designated by the letters DN followed by a number

❑ ***Pressure Equipment Regulations 1999***

• NON-IONISING RADIATION

Radiation caused by exposure to lasers, ultra-violet, infra-red, microwave and radio frequencies

• NORMAL DAY TO DAY ACTIVITIES

Normal day to day activities are defined as the ability of a person to carry out any of the following activities

- mobility
- manual dexterity
- physical co-ordination
- continence
- ability to lift, carry or otherwise move everyday objects
- speech
- hearing or eyesight
- memory or ability to concentrate, learn or understand
- perception of risk of physical danger.

❑ ***Disability Discrimination Act 1995***

• NOTIFIABLE

A number of activities or conditions are notifiable to various authorities or agencies, for example:

- accidents and diseases under RIDDOR
- cooling towers
- installations handling hazardous substances
- new substances.

• NOTIFIED BODY

A body appointed in the UK to carry out conformity assessment procedures in relation to lifts, e.g. examinations, production checks and quality assurance systems

❑ ***Lifts Regulations 1997***

• NRPB

The abbreviation for the National Radiological Protection Board

• NUCLEAR WASTE

Radioactive waste produced as a by-product of the nuclear industry

• NUISANCE

A nuisance is an act which interferes with the normal enjoyment of a right

Everyone has a right to enjoy peace and quiet, but being disturbed by noise is a nuisance. The noise becomes a statutory nuisance when the noise exceeds a certain level.

• OBITER DICTA

Obiter dicta is the Latin phrase meaning 'other things said' or words said 'by the way'

In reaching his or her verdict, the judge's speech contains the legal principles on which the decision is based. This is called the *ratio decidendi* and forms the binding precedent. Anything else said by the judge is *obiter dicta* and does not form part of the binding precedent although it may be termed persuasive.

• OBJECTIVITY

A concept used in connection with determining reasonableness both in criminal and civil cases where the court may pose the question – 'would an ordinary person have realised the consequence of such an act?'

If the answer is in the affirmative, a defendant would be deemed liable even if he or she had not realised the consequences. This is also referred to as the objective test.

• OCCUPATIONAL DISEASE

A disease caused by exposure to certain occupational hazards

❑ ***Social Security (Industrial Injuries) (Prescribed Diseases) Regulations 1985***

• OCCUPATIONAL EXPOSURE STANDARD

The concentration of an airborne substance averaged over a reference period at which there is no current evidence that repeated exposure by inhalation will affect an employee's health

➔ *MEL*

❑ ***EH40***

• OCCUPIER

A person who has sufficient degree of control over premises or part of premises to owe a *duty of care* to other persons having lawful access, e.g. employees, visitors, etc.

• OCCUPIER'S LIABILITY

The *duty of care* owed by the occupiers of premises to visitors and to the premises

❑ ***Occupiers' Liability Acts 1957 and 1984***

• OFFER

An expression by one person that he (or she) is willing to contract with another person or persons on specified terms

If it is to form the basis of a contract, the person making the offer must intend that legal consequences shall result. An offer can be made verbally or in writing. ➔ *invitation to treat*

• OFFICIOUS BYSTANDER TEST

This is a test used in court when contract disputes are being heard and there is argument concerning an implied term

The test applied is whether an officious bystander, on hearing both parties discussing the terms of the contract, would have pointed out to them that a term should be included and both parties would have agreed that its inclusion was so obvious as to be unnecessary. The court would take the view that such a term would be implied in the contract.

• OHM

An ohm (O) is a measure of electrical resistance to current

• OHM'S LAW

Ohm's Law states that the current in a circuit is directly proportional to the voltage and inversely proportional to the resistance providing the temperature remains constant

This can be expressed in a number of ways:

- voltage = current times resistance
- volts = amps x ohms
- V = A x **Ω**

• OMBUDSMAN

A person who has been appointed to deal with complaints, such as delay and inefficiency, relating to certain areas such as banking or health, or local government

• ORGANIC WASTE

Waste of biological origin derived from oil, coal or living organisms

• OXYGEN ENRICHMENT

Oxygen enrichment occurs when the normal oxygen content of the air in a confined space is supplemented by the use of oxygen supplied from an oxygen cylinder

This is hazardous and can lead to fire and/or explosion.

• OZONE

A colourless gas at room temperature which is minutely present in the atmosphere

It is generated during welding and can also be produced by photochemical reaction. Photocopiers can produce ozone in small quantities. Ozone is an irritant which affects the nose and throat and can impair breathing in persons with heart or lung conditions.

• OZONE LAYER

A band of ozone encircling the earth which protects living organisms from the sun's ultra-violet radiation

At the South Pole, there is a hole in the ozone layer which expands and contracts naturally according to the seasons.

In recent years, it has been discovered that the hole is getting larger and there is some disagreement in the scientific world as to the cause with some scientists believing it to be a natural phenomenon and some attributing it to human activity mainly because of the use of CFCs. ➔ *Montreal Protocol*

• PARALLAX

A parallax view is the phenomenon which occurs when an object is looked at from two different places

The object appears to change position in relation to its background.

• PARASITE

An organism which lives off another organism

• PARETO PRINCIPLE

A principle propounded by economist Vilfredo Pareto in the 1800s which has become widely used in management training to illustrate the need to focus on essentials

In many groupings, it has been found that there is an 80:20 per cent split – e.g. in a working group perhaps 80 per cent of the production is produced by 20 per cent of the workers.

• PASCAL

A unit of pressure equal to one Newton per square metre

• PASSIVE SMOKING

The inhalation of smoke from cigarette and pipe smoke by non-smokers in the vicinity of a smoker

It has become a contentious problem in the workplace and in other public places as there are fears that lung problems can be caused by passive smoking.

• PASTEURISATION

A method of killing bacterial cells by means of heat treatment

• PATERNOSTER LIFT

A continuous-running, slow-moving lift which has no front door or gate

Passengers step onto and from the slow-moving platform at their own risk.

• PATHOGEN

A micro-organism that may cause illness

• PEARSON COMMISSION

The Royal Commission on Civil Liability and Compensation set up in 1978

It examined the system of providing compensation for persons who had suffered personal injury as a result of accidents at work. It recommended that the current system of compensation by a mixture of social security benefit and common law damages should be retained.

• PEER GROUP PRESSURE

The pressure exerted by being a member of a particular group of individuals of equal status where the individual member is strongly influenced by group culture, behaviour and expectations

• PERCEPTION

The reception and interpretation of information

Individual perception is influenced by such factors as personal experience, education, training, intelligence, emotional state, etc. It is important in safety terms from the differing personal interpretations of what is dangerous

• PERCEPTUAL SET

The combination of components or factors by which an individual interprets and makes sense of the available data which surrounds him or her

• PERMANENT JOINTS

Joints which cannot be disconnected except by destructive methods

• PERMANENT THRESHOLD SHIFT

Permanent damage to hearing as a result of exposure to noise

• PERMIT TO WORK

A permit-to-work system is a formal written authorisation permitting certain hazardous work to be carried out by specified competent persons

It indicates the work to be undertaken; the safety precautions involved; the safety hazards identified; and the risk assessments that have been carried out.

The permit is issued by an authorised person to the person who is to carry out the work specifying;

- the work to be carried out and by whom
- the time-scale in which it is to be carried out
- the safety measures required and the necessary precautions to be taken.

A copy of the permit must be signed and retained by the person authorising it, the person carrying out the work, and the person in charge of the area where the work is to be carried out. When the work is complete, the permit is returned to the authorised person and cancelled.

• PERSONAL INJURY

Personal injury includes loss of life, any impairment of a person's physical or mental condition and any disease

❑ *Employer's Liability (Defective Equipment) Act 1969*

• PERSONAL PROTECTIVE EQUIPMENT (PPE)

PPE means all equipment (including clothing affording protection against the weather) which is intended to be worn or held by a person at work and which protects him against one or more risks to his health or safety, and any addition or accessory designed to meet that objective

❑ *Personal Protective Equipment Regulations 1992/1994*

• PERSONAL SUSPENSION EQUIPMENT

Suspended access equipment (other than a working platform) for use by an individual, including a boatswain's chair and abseiling equipment, which must be of suitable and sufficient strength for its load

It must be securely attached to a structure which is of sufficient strength to support the load and installed in such a way as to prevent any uncontrolled movement.

❑ *Construction (Health, Safety and Welfare) Regulations 1996*

• PERSUASIVE PRECEDENTS

Statements and decisions of courts which are not legally binding on other courts but which might 'persuade' those other courts to follow or be influenced by the statement or decision.

➔ *precedent*

• PETROLEUM

Any mineral oil or related hydrocarbon and natural gas petroleum existing in natural condition in strata

It includes oil made from petroleum, coal, shale, peat or other bituminous substances

• PETROLEUM MIXTURES

Substances containing petroleum and having a flashpoint under 73°F. They include paint strippers, cleaning solvents, cellulose, cellulose thinners and rubber adhesive solutions

• PETROLEUM SPIRIT

Petroleum spirit includes benzine, petrol, petroleum, petroleum naphtha, toluene and xylene each having a flash point under 73°F

• pH

A measure of the acidity.or alkalinity of a substance

On a scale of 0 to 14 – figures below (approximately) 7 are acidic and those from 7 to 14 alkaline, e.g. water has a pH of 7.1 – a neutral solution.

• PHAGOCYTE

Part of the body's immune system

It is a cell found in various parts of the body – blood, spleen etc., which is capable of surrounding and engulfing bacteria which have entered the body.

• PHOTOCOPIER

A photocopier can be a largely ignored source of health and safety hazards

Ultra-violet light and ozone emission which can cause headaches and sickness often attributed to sick-building syndrome. Replenishing toners, which should be subject to COSHH assessments, can also lead to health problems.

• PHOTOSYNTHESIS

The process by which plants use light to convert water and carbon dioxide into sugar

• PIPELINE

A pipe or system of pipes used for the conveyance of relevant fluid across the boundaries of premises, together with any

apparatus for inducing or facilitating the flow of relevant fluid through, or through a part of, the pipe or system, and any valves, valve chambers, pumps, compressors and similar works which are annexed to, or incorporated in the course of, the pipe or system

❑ ***Pressure Systems Safety Regulations 2000***

• PIPER ALPHA DISASTER

An explosion on an oil rig in the North Sea in 1988 which caused the loss of 167 lives

The subsequent enquiry into the causes of the disaster culminated in the *Cullen Report* and the *Safety Case Regulations*.

• PIPING

Piping components intended for the transport of fluids when connected together for integration into a pressure system

Such components include in particular a pipe or system of pipes, tubing, fittings, expansion joints, hoses, other pressure-bearing components as appropriate or heat exchangers consisting of pipes for the purpose of cooling or heating air.

❑ ***Pressure Equipment Regulations 1999***

• PLACE OF SAFETY

A place in which persons are not in any danger from fire

❑ ***BS4422***

• PLAINTIFF

A person presenting a claim in a civil action

• PLANNED MAINTENANCE PROGRAMME

An organised maintenance schedule based on pre-planning, previous maintenance and service records, frequency, time intervals, etc.

• PLANNING SUPERVISOR

A Planning supervisor under the CDM Regulations is a company, partnership, organisation or an individual who co-ordinates and manages the health and safety aspects of the design of the construction project

The planning supervisor also has to ensure that the pre-tender stage of the *health and safety plan* and the *health and safety file* are prepared.

• PLATFORM

Platforms from which a person could fall more than 2m must be fitted with guard rails and toe boards

This also applies to intermediate platforms. Guardrails must be 910mm above the platform with the toe boards at least 150mm high with no more than 765mm between the top of the toe board and the guard rail. ➔ *scaffolding*

• POLICY

The safety policy is the key document in the management of health and safety which specifies how each individual organisation is to fulfil its obligations under the HSWA 1974

'Except in such cases as may be prescribed, it shall be the duty of every employer to prepare and, as often as may be appropriate, revise a written statement of his general policy with respect to the health and safety at work of his employees and the organisation and arrangements for the time being in force for carrying out that policy, and to bring the statement and any revision of it to the notice of all his employees'

A written policy is required where there are five or more employees.

❑ ***Health and Safety at Work etc Act 1974***

• POLING BOARDS

Boards fixed vertically to support the sides of excavations

• POLLUTANT

Any organism or substance which contaminates the purity of air, water, or soil

• POLLUTION

The destruction of natural environment by a contaminant

• POLYCHLOROPRENE

A plastic compound used to provide protective sheathing for electric cables

• POLYMER

A substance comprising a simple weight majority of molecules containing at least 3 monomer units which are covalently bound to at least one other monomer unit or other reactant

• POLYVINYL CHLORIDE (PVC)

PVC is the abbreviation for polyvinyl chloride the most commonly-used plastic

PVC is used extensively as protective sheathing for electric circuits. It is dangerous in a fire situation as it releases hydrochloric acid gas during burning.

• POST-TRAUMATIC STRESS DISORDER

A direct response to a specific stressful event, e.g. the stress experienced by persons involved in serious accidents or major disasters

➔ *stress, trauma*

• PORTABLE APPLIANCE TESTING

PAT is used to test the safety of portable electrical appliances particularly with regard to insulation and earthing

• PORTABLE ELECTRICAL EQUIPMENT

Any item of electrical apparatus which uses mains electrical power and which is capable of being carried and used by hand

The term covers items such as power tools, hand lamps, etc. but any electrical equipment which can be unplugged and transported would come within the definition.

• PORTABLE FIRE EXTINGUISHERS

➔ *classification of fire*

• PRACTICABLE

➔ *so far as is reasonably practicable*

• PRACTICAL

➔ *so far as is reasonably practical*

• PRECEDENT

Precedent occurs when a judge makes a decision based on the facts of a particular case

Decisions of higher courts are regarded as forming precedents which are binding on lower courts when considering cases based on similar facts. Once a decision of a higher court has been given the status of a precedent, it must be followed by a lower court where the same point of law is involved. A precedent is binding on a lower court. It is this use of judicial precedent which gives rise to case law. ➔ *judicial precedent*

• PREPARATIONS

Mixtures or solutions of two or more substances

❑ ***Notification of New Substances Regulations 1993***

• PRESBYCUSIS

The progressive loss of hearing associated with ageing

It can be exacerbated by continual exposure to high noise levels.

• PRESCRIBED OCCUPATIONAL DISEASE

An industrial disease which provides an entitlement to financial benefit for those claimants who can prove that they have incurred the disease as a result of exposure to certain occupational conditions

➔ *occupational disease*

• PRESSURE

Pressure means pressure relative to atmospheric pressure being gauge pressure; vacuum is designated by a negative value

❑ ***Pressure Equipment Regulations 1999***

• PRESSURE EQUIPMENT

This means vessels, piping, safety accessories and pressure accessories

Where applicable, pressure equipment include elements attached to pressurised parts, such as flanges, nozzles, couplings, supports, lifting lugs, and similar.

❑ ***Pressure Equipment Regulations 1999***

• PRESSURE SYSTEM

A system comprising one or more pressure vessels of rigid construction, any pipe-work and protective devices

This definition includes the pipe-work with its protective devices to which a transportable gas container is, or is intended to be connected; a pipeline and its protective devices, which contains or is liable to contain a relevant fluid, but does not include a transportable gas container. The word 'liable' in this last definition means that systems which only occasionally contain a relevant fluid are, nonetheless, defined as 'pressure systems' and subject to the requirements of the regulations.

❑ ***Pressure Systems and Transportable Gas Containers Regulations 1989***

• PRINCIPAL CONTRACTOR

The main contractor appointed by a *client* under the CDM Regulations who has overall responsibility for the management of site operations, including overall co-ordination of site health and safety management

❑ ***Construction (Design and Management) Regulations 1994***

• PRIVILEGE

The term privilege is applied in court procedure when certain evidential documents are deemed to be privileged and need not be disclosed to the opposition

• PROBABILITY

The likelihood of an event occurring

Probability theory is used in health and safety to try to determine the chance of an incident occurring and the degree of damage or injury likely to ensue.
➔ *risk*

• PROCESS-ORIENTATED RESEARCH AND DEVELOPMENT

The further development of a substance in the course of which pilot plant or production trials are used to test the fields of application of the substance

• PROHIBITION NOTICE

A formal notice served by a HSE Inspector on an employer requiring him to cease, immediately, any procedure which, in the inspector's opinion, will involve a risk of serious personal injury to any person

➔ *improvement notice*

❑ ***Health and Safety at Work etc Act 1974***

• PROHIBITION SIGN

A sign prohibiting behaviour likely to cause a risk to health or safety

❑ ***Health and Safety (Safety Signs and Signals) Regulations 1996***

• PROPANE

A colourless, gaseous hydrocarbon

In liquid form it is used as a fuel.

• PROSECUTION

Instituting and carrying on a criminal charge or other legal proceedings in court

Prosecution of breaches of health and safety legislation may only be undertaken by officers appointed by the enforcing authorities (the inspectors), or the Director of Public Prosecutions, or someone acting under his instructions.

Where proceedings are brought before a Magistrate's Court, the inspector may prosecute without being legally qualified. In the case of a hearing before a Crown Court, an inspector is not permitted to prosecute the case. The Director of Public Prosecutions may, in any court, consent to proceedings being instituted by anyone, thus a private prosecution by an individual is possible provided that consent is obtained.

❑ ***Health and Safety at Work etc Act 1974***

• PROTECTIVE CLOTHING

Clothing specifically worn for protection of health and safety

This includes aprons, protective clothing for adverse weather conditions, gloves, safety footwear, safety helmets, and high visibility waistcoats. It does not include ordinary working clothes and uniforms which do not specifically protect the health and safety of the wearer.

❑ ***Personal Protective Equipment Regulations 1992***

• PROTECTIVE DEVICES

Devices designed to protect the pressure system against system failure and devices designed to give warning that system failure might occur, and include bursting discs

• PURGING

Cleaning out or flushing a container or pipe system of unwanted impurities or gases

• PUTLOG

A transverse horizontal metal tube with a flattened end which is inserted into new brickwork with the other end fixed to a standard

➔ *scaffold* and *transom*

• PUTLOG SCAFFOLD

One with a single row of standards – vertical scaffold tubes or poles

➔ *scaffold*

• PYROLYSIS

The breaking up of a substance by heat

• PYROPHORIC

A pyrophoric substance is one which ignites readily on exposure to air

• QUALIFIED MAJORITY VOTING

A voting system operating within the Member States of the EU in order to expedite the passing of legislation without the necessity of unanimity amongst all the States

Individual States have an allocated number of votes, totalling in all 87. Of the total number of votes available, 62 votes amount to a qualified majority. This will change with the admission of new member states in the future.

• QUALITATIVE

The term 'qualitative' is applied, in managing health and safety, where the researcher or quality control manager gathers information through the techniques of observation and interview and uses the information thus obtained to make a subjective judgment based on personal experience

➔ *quantitative*

• QUANTITATIVE

The term 'quantitative' in a health and safety context is used where the researcher obtains information from established research or statistics or by means of structured questionnaire or observation

The numerical data or information thus obtained may be used as a basis for risk assessment or probability studies

• QUARTZOID BULB

A small glass bulb in a sprinkler head which holds the water valve in the sprinkler in place

The bulb contains a liquid which expands, as the temperature increases in a fire situation, until it shatters the bulb releasing the water valve and allowing the sprinkler to operate. By adjusting the composition of the liquid inside it, the bulb can be set to operate at a desired temperature usually within the range 57º – 260ºC. The bulbs are colour coded according to the temperature rating.
➔ *sprinkler system*

• RACIAL DISCRIMINATION

Direct racial discrimination occurs where on racial grounds a person is treated less favourably than others would be treated or that person is segregated from others

Indirect racial discrimination occurs where, for example, in employment, there appears to be equality of treatment where a requirement or condition is applied to a large group of people. However, the number of persons in a particular racial group who are able to comply with the requirement is proportionately smaller than the number of persons outside the racial group who can comply and the employer cannot justify the requirement as being necessary regardless of racial origin, and, being unable to comply puts the person at a disadvantage.

❑ ***Race Relations Act 1976***

• RADIATION

The transmission of light, sound or heat by 'invisible rays' emanating from an object

The rays can travel through air and through a vacuum but are stopped by solids and liquids. Radiation travels in straight lines at the speed of light. The heating effect diminishes with distance. If the distance between an object and a source of radiation is doubled, the amount of radiation received decreases to one quarter.

Transparent materials allow the rays to pass through without any heating effect. Many other materials either absorb the radiation and become hot or reflect it. The best reflectors are shiny or white, smooth objects. Blackened objects do not reflect well but radiate heat more easily. Radiation is ionising or non-ionising.

• RADIATION ACCIDENT

This means an accident where immediate action would be required to prevent or reduce the exposure to ionising radiation of employees or any other persons

• RADIATION EMPLOYER

An employer who in the course of a trade, business or other undertaking carries out or intends to carry out work with ionising radiation

• RADIONUCLIDE

A disintegrating nucleus of an atom which releases radioactivity

• RADIATION PROTECTION ADVISER

An individual who, or a body which, meets such criteria of competence as may from time to time be specified in writing by the Executive

❑ ***Ionising Radiations Regulations 1999***

• RADIOACTIVE SUBSTANCE

Any substance which contains one or more radionuclides whose activity cannot be disregarded for the purposes of radiation protection

❑ ***Ionising Radiations Regulations 1999***

• RADIATION SICKNESS

Radiation sickness is caused by exposure to a dose of ionising radiation between 100-300 rem

Exposure leads to nausea, damage to the bone marrow and can cause leukaemia.

• RADON GAS

A naturally occurring, invisible, odourless gas produced as a result of radioactive decay of uranium in soil and rock

It is radioactive and thought to be a carcinogenic.

• RATIO DECIDENDI

The Latin phrase for the 'reason for the decision' and is the part of the judge's judgement which forms the binding precedent of the case

➔ *judicial precedent*

• REACH-TRUCK

A truck where the lifting forks are accommodated within the

wheelbase of the truck while the load is being carried and projected forward by moving the mast and forks forward when the load is to be raised or deposited

This type of truck is not counterbalanced and is more manoeuvrable when carrying a load.

➔ *fork-lift*

• REDUCED VOLTAGE

A reduced voltage system is one where the voltage is reduced in order to prevent severe electric shock

A voltage of 55v for users of portable electric equipment which will prevent an operator from receiving a fatal shock can be achieved witha 110v transformer *centre-tapped* to earth.

• RELEVANT AGREEMENT

A workforce agreement which applies to a worker, any provision of a collective agreement which forms part of a contract between him and his employer, or any other agreement in writing which is legally enforceable as between the worker and his employer

❑ ***Working Time Regulations 1998***

• RELEVANT FLUID

The regulations define 'relevant fluid' to mean either (a) steam; any fluid or mixture of fluids which is at a pressure greater than 0.5 bar above atmosphere and is either a gas or a liquid which would have a vapour pressure greater than 0.5 bar when in equilibrium with its vapour at either the actual temperature of the liquid or 17.5oC; or (b) a gas dissolved under pressure in a solvent contained in a porous substance at ambient temperature and which could be released from the solvent without application of heat, e.g. acetylene

❑ ***Pressure Systems and Transportable Gas Containers Regulations 1989***

• RELEVANT STATUTORY PROVISIONS

Originally a number of existing statutory provisions which were contained in Schedule 1 of the HSWA 1974 which were to be regarded as part of the Act pending their progressive replacement

• RELEVANT TRAINING

Work experience provided pursuant to a training course or programme, training for employment, or both, other than work experience or training, the immediate provider of which is an education institution or a person whose main business is the provision of training, and which is provided on a course run by that institution or person

❑ ***Working Time Regulations 1998***

• RELEVANT TRANSPORT SYSTEM

A railway, tramway, trolley vehicle system or guided transport system

❑ ***Reporting of Injuries, Diseases and Dangerous Occurrences Regulations 1995***

• REM

The acronym for Roentgen Equivalent Man which was a measure of the biological effect of radiation on the body

➔ *Sievert*

• REMOTENESS OF DAMAGE

A term used in negligence cases

A defendant is not liable for damage that is not reasonably foreseeable. Such damage is said to be remote.

• REPEAL

The legal term for the cancellation or annulment of primary legislation, e.g. a statute or part of a statute

➔ *revoke*

• REPETITIVE STRAIN INJURY (RSI)

A physical condition usually affecting hands and arms arising from continued repetition of physical movements

➔ *work-related upper limb disorder*

• RESIDUAL CURRENT DEVICE

A mechanical switching device which is designed to break the contacts in an electrical circuit when it senses a leakage of current to earth

The given value (leakage) is usually set at 30mA and the trip mechanism breaks the circuit within 30m seconds thus preventing electric shock.

➔ *ELCD*

• RES IPSA LOQUITUR

A Latin phrase meaning 'the thing speaks for itself'

It occurs in legal cases when the circumstances of an accident are such that the negligent behaviour which caused the accident points inevitably to the accused. In such cases, the burden of proof shifts to the defendant to disprove the negligence.

• RESPIRABLE DUST

Airborne material which is capable of penetrating to the gas exchange region of the lung

• RESPIRATORY PROTECTIVE EQUIPMENT (RPE)

RPE is breathing equipment for use in dangerous atmospheric conditions

The respirators range from simple face masks for protection against respirable particles (see *respirable dust*) to self-contained breathing apparatus with its own air or oxygen supply. RPE must comply with PPE Regulations and be approved by the HSE. Provison of RPE is a last resort measure which must only be introduced after all other methods of protection or avoidance have been considered

• RESPIRATORY SYSTEM

In simple terms, the combination of the mouth, nose and nasal passages, throat and lungs

• RESPONDENT

A person against whom an appeal is brought in a civil action

• RESPONSIBLE PERSON

The person for the time being in control of premises in which he or she carries on any trade or business in or at which, the reportable accident, dangerous occurrence or reportable disease occurred

❑ ***Reporting of Injuries, Diseases and Dangerous Occurrences Regulations 1995***

• REST PERIOD

A period which is not working time, other than a rest break or leave, to which the worker is entitled under these Regulations

❑ ***Working Time Regulations 1998***

• REVEAL TIE

A scaffolding tube used to 'tie' the scaffold to a building

The reveal tie is wedged between the opposing sides of an aperture (doorway, window etc) by means of an adjustable pin at one end of the tube. It is then coupled to a tube connecting it to the scaffold. ➔ *through tie*

• REVOKE

The legal term used for the cancellation or annulment of subordinate legislation, e.g. a regulation, statutory instrument or part thereof

In short, Acts of Parliament are repealed – Regulations are revoked.

➔ *repeal.*

• RIDDOR

The acronym for the Reporting of Injuries, Diseases and Dangerous Occurrences Regulations

Under the Regulations, whenever any of the following incidents occur 'out of or in connection with work', it must be reported to the enforcing authority by the quickest practicable means (e.g.telephone) and in writing within ten days, and a record kept:

- the death of any person as a result of an accident, whether or not he or she is at work
- a person who is at work suffers a major injury as a result of an accident
- a person who is at not at work (e.g. a member of the public) suffers an injury arising out of or in connection with work and is taken from the scene to a hospital for treatment in respect of that injury
- a person not at work suffers a major injury as a result of an accident arising out of or in connection with work at a hospital
- one of a list of specified 'dangerous occurrences' takes place (dangerous occurrences are events which do not necessarily result in a reportable injury, but have the potential to do significant harm)
- where a person at work is unable to do his or her normal work for

more than three days as a result of an injury (an 'over-3-day' injury) caused by an accident at work.

❑ ***Reporting of Injuries, Diseases and Dangerous Occurrences 1995***

• RISK

The chance or likelihood that harm from a particular hazard might be realised

• RISK ASSESSMENT

An evaluation of the hazards and risks involved in a particular process or activity and a judgement as to how these can be eradicated or reduced to an acceptable level

In order to comply with legal requirements, which require a risk assessment to be suitable and sufficient, the following points must be addressed.

- Identify all the hazards and evaluate the risks arising from them.
- Record significant findings (if more than 5 employees).
- Identify employees who are most at risk.
- Identify others who might be affected, e.g. visitors, members of the public, etc.
- Evaluate existing controls (whether they are satisfactory or not).
- Judge and record the probability of an accident occurring describing the 'worst case' outcome.
- Identify the information needed for employees, the risks, the precautions and any emergency arrangements.
- Prepare an action plan of preventive measures.
- Incorporate any significant change in process or activity (reassess).

Where work equipment and work tasks are common throughout a workplace, a generic assessment may be carried out on one common piece of equipment or task which can then be applied to all the others.

The following separate pieces of legislation require risk assessment in one form or another:

❑ ***The Management of Health and Safety at Work Regulations 1994***
❑ ***The Manual Handling Operations Regulations 1992***
❑ ***The Personal Protective Equipment at Work Regulations 1992***
❑ ***The Health and Safety (Display Screen Equipment) Regulations 1992***
❑ ***The Noise at Work Regulations 1989***
❑ ***The Control of Substances Hazardous to Health Regulations 1994***
❑ ***The Control of Asbestos at Work Regulations 1980***
❑ ***The Control of Lead at Work Regulations 1980***

❑ ***The Construction (Design and Management) Regulations 1994***
❑ ***The Construction (Health, Safety and Welfare) Regulations 1996***

• RISK ASSESSMENT – 5 STEPS

The HSE has issued advice on risk assessment known as the '5 Steps'

The 5 steps are:

Step 1	Identify the hazard
Step 2	Identify who might be harmed
Step 3	What more is needed to control the risk
Step 4	Record the assessment
Step 5	Review and revise the assessment

• RISK ASSESSMENT EVALUATION

A method of determining the risk factor involved in an activity by assigning a numerical weighting to the categories of severity and probability

A simple matrix showing a 'probability' scale horizontally and a 'severity' scale vertically with an appropriate numerical weighting, e.g. 25 to 1, will produce a numerical weighting which can be used as a rough guide to the risk factor involved in an activity.

		Certain	Very likely	Likely	Unlikely	Very unlikely
		5	4	3	2	1
Fatality	5	25	20	15	10	5
Major injury Disease Disabled	4	20	16	12	8	4
Minor injury	3	15	12	9	6	3
Damage	2	10	8	6	4	2
Delay	1	5	4	3	2	1

A simple multiplication of the two figures will give a risk factor ranging from 25 – 1, e.g. a certain fatality would score 25 and a likely minor injury 9.

25-15 risk unacceptable or not adequately controlled and requiring immediate or urgent action

10-12 compare with current standards and arrange for appropriate controls to be instituted

5 – 9 adequately controlled risk

1-4 trivial risk

This rough rule of thumb allows a judgment to be made in order to prioritise and determine the value of any action considered to be 'Reasonably Practicable'.

• RISK AVOIDANCE

Risk avoidance is when an organisation decides to completely avoid a risk by discontinuing a hazardous operation

• RISK MANAGEMENT

The process of turning a risk into a manageable problem

This involves analysing, measuring and distributing risks in order to reduce their potential for damage to a manageable and acceptable level.

➔ *risk avoidance, risk reduction, risk retention, risk transfer.*

• RISK PHRASE

A risk phrase relating to dangerous substances listed in the *approved supply list*

In these Regulations, specific risk phrases may be designated by the letter 'R' followed by a distinguishing number or combination of numbers. The risk phrase must be quoted in full on any label or safety data sheet on which the risk phrase is required to be shown, e.g., R25 – toxic if swallowed; R43 – may cause sensitisation by skin contact, etc.

➔ *safety phrase* and *Approved Supply List*

• RISK RATING

A risk rating is achieved by multiplying together the assigned numerical values of probability; and severity

➔ *risk assessment evaluation*

• RISK REDUCTION

When an organisation decides to address the problem of risk by introducing a loss control programme

• RISK RETENTION

When an organisation decides to 'retain' a known risk and cover any resultant financial loss from within its own resources

• RISK TRANSFER

When an organisation decides to transfer the costs of a potential risk to another party, usually by way of insurance

• ROOFS

Roof structures themselves do not always provide a safe handhold or foothold for workers. For work on sloping roofs or work near the edge of flat roofs, there must be sufficient edge protection to prevent falls of materials and persons.

Where work is being carried out on a sloping roof of more than 30° (or less than 30° but which is slippery) crawling ladders or crawling boards must be used.

Fragile roofs require particular care and where work is carried out on or near fragile materials such as asbestos cement sheets or glass, crawling boards must be used and warning notices posted All roof lights must be properly covered or provided with barriers. Where other workers are working under roof work, precautions must be taken to prevent debris falling on them.

• ROOM DIMENSIONS

Room dimensions where people work must provide adequate space – floor area, height and space – for the purpose of health, safety and welfare

According to the relevant Code of Practice, each person must have a minimum of 11 cubic metres of space calculated by dividing the volume of the room by the number of persons who normally work in it. In arriving at the calculation, any room with a height greater than 3 metres should be deemed to be only 3 metres in height.

❑ ***Workplace (Health, Safety and Welfare)) Regulations 1992***

• ROSPA

The acronym for the Royal Society for the Prevention of Accidents

• ROUTES OF ENTRY

The various routes through which hazardous substances may enter the body, i.e. *inhalation, ingestion, absorption* and *direct entry*

• ROYAL ASSENT

The final part of the legislative process when the sovereign signs the legislation making it an Act of Parliament

• SAFE

Safe in relation to pressure equipment or an assembly, means that the pressure equipment or assembly when properly installed and maintained and used for its intended purpose is not liable to endanger the health or safety of persons and, where appropriate, domestic animals or property

• SAFE OPERATING LIMITS

The operating limits (incorporating a suitable margin of safety) beyond which system failure is liable to occur

• SAFE PLACE OF WORK

The provision of a Safe Place of Work for employees is an obligation placed on employers under the HSWA 1974

• SAFE SYSTEMS OF WORK

A safe system of work is a combination of a number of factors – the layout of the workplace, the order in which tasks are carried out, and any special instructions or precautions in relation to any specific hazardous operation

A safe system of working is a method of carrying out an operation in a manner which is safe and without risk to the operator and anyone who might be affected by the performance of the activity. Safe systems are achieved through efficient risk assessments.

• SAFE WORKING LOAD

The maximum weight that any mechanical lifting equipment can handle in safety, e.g. lifts, cranes, trucks, etc. The 'SWL' must be clearly marked on any such equipment

• SAFETY ACCESSORIES

Devices designed to protect pressure equipment against the allowable limits being exceeded

These include:

- devices for direct pressure limitation, such as safety valves, bursting disc safety devices, buckling rods, controlled safety pressure relief systems, and
- limiting devices, which either activate the means for correction or provide for shutdown or shutdown and lockout. such as pressure switches or temperature switches or fluid level switches and safety related measurement control and regulation devices

❑ ***Pressure Equipment Regulations 1999***

• SAFETY ADVISER

A person appointed by an employer under the MHSW Regulations to advise and assist him or her in health and safety matters affecting the organisation

He or she must be competent and may be an employee or an independent consultant. The term also occurs in relation to the transport of dangerous goods where an employer must appoint a safety adviser to advise on environmental, health and safety matters in relation to their carriage and transport.

❑ ***Management of Health and Safety at Work Regulations 1992***

• SAFETY AUDIT

The systematic measurement and validation of an organisation's management of its health and safety programme against a series of specific and attainable standards

Every component of the total system is included, e.g. management policy, attitudes, training, features of processes, personal protection needs, emergency procedures, etc. This is followed by recommendations leading to achievable targets being set prior to each subsequent audit.

➔ *safety monitoring*

• SAFETY CASE

A safety case, in relation to offshore installations, is a written specification indicating that there is an adequate management system to ensure that all statutory regulations are complied with;

that there are adequate arrangements to audit and report on the management system; that all major hazards have been identified; and that all risks have been assessed and measures are in force to keep risks to the lowest reasonably practicable level

→ Cullen Report

❑ ***Safety Case Regulations 1992***

• SAFETY COLOUR

Safety Colour means a colour to which a meaning is assigned, e.g. safety signs are colour-coded as follows:

Red	Prohibition Sign
Blue	Mandatory Sign
Yellow	Warning Sign
Green	Information Sign

❑ ***Health and Safety (Safety Signs and Signals) Regulation 1996***

• SAFETY COMMITTEE

A committee composed of shop-floor and management employees to consider and make recommendations on health and safety matters

An employer must appoint a safety committee where he or she is requested to do so by two or more trade union safety representatives.

❑ ***Safety Representatives and Safety Committees Regulations 1977***

• SAFETY CONSULTANT

An independent consultant who provides advice on aspects of health and safety

He or she should hold appropriate health and safety qualifications and be experienced in the specific field in which advice is being sought.

❑ ***INDG133 – Selecting a Health and Safety Consultant***

• SAFETY CULTURE

A system of shared values and beliefs amongst management and staff about the importance of health and safety in the workplace

According to the HSE, *'the safety culture of an organisation is the product of individual and group values, attitudes, perceptions, competencies and patterns of behaviour that determine the commitment to, and the style and proficiency of, an organisation's health and safety management.'*

They suggest that the best way to ensure good health and safety practices within an organisation is to develop a 'positive safety culture' and that such development lies in the hands of senior management and involves the four Cs:

- **Competence** – identification and development of skills resulting in a workforce which is well-informed and knowledgeable about risks and the necessary precautions and procedures for controlling them
- **Control** – clear demonstration of commitment – an organisational structure in which responsibilities are clearly defined and people are accountable
- **Co-operation** – the extent to which employees are involved in planning/developing safe systems of work and monitoring performance – an atmosphere in which all are actively involved in continuous improvement
- **Communication** – provision of information about risks, plans, objectives and feed-back on performance. An atmosphere in which individuals are encouraged to report hazards and near misses as well as injuries

❑ ***HSG65***

• SAFETY DATA SHEET

A document provided to customers by suppliers of dangerous substances

It contains:

- name of supplier
- name of substance
- ingredients: physical and chemical properties
- toxicity
- hazard identification
- first-aid requirements
- fire-fighting requirements
- accidental release measures
- handling and storage
- PPE requirements
- disposal procedure and ecological effects
- transportation requirements.

The safety data sheets must be current and in the language of the country of the intended customer.

• SAFETY GLASS

Safety glazing materials must be used in doors, windows, walls, etc where the width of the aperture is more than 250mm

Safety glazing materials include:

- glass blocks
- safety glass
- Georgian wired glass
- annealed glass
- polycarbonate material.

• SAFETY INSPECTIONS

Generally, a scheduled inspection of premises or departments by personnel within that organisation, or by an external specialist

This is in order to:

- examine maintenance standards, employee involvement, working practices and housekeeping levels
- to check that work is undertaken in accordance with company codes of practice or procedures
- to ensure that specific procedures designed to promote safe working are being operated
- to protect their employers from liability.

Trade union safety representatives carry out inspections to protect members of their trade union from hazards. Inspectors of the enforcing authorities inspect in order to identify breaches of the law regarding the causes of a notified accident.

➔ *safety monitoring*

• SAFETY MANAGEMENT

The HSE bases its safety management system on 5 stages – policy – organising – planning – measuring performance – and reviewing performance

It is usually represented by the following diagram:

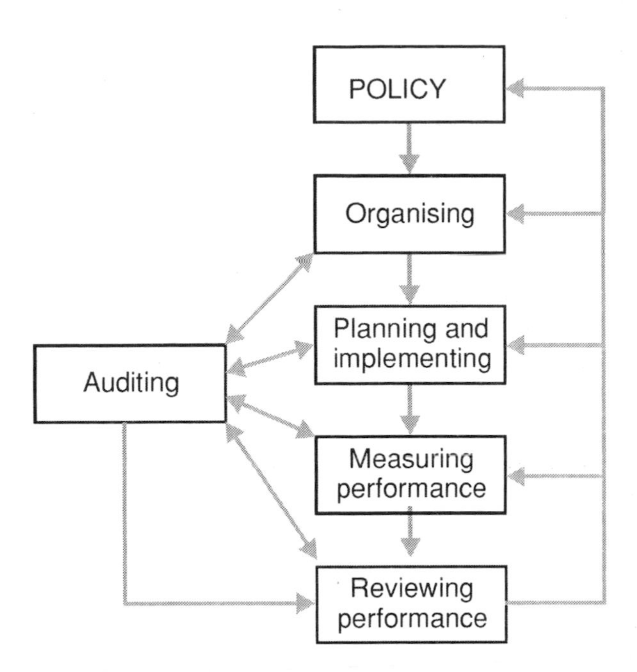

❑ ***Successful Health & Safety Management HSG65***

• SAFETY MONITORING

Health and safety monitoring covers all of the activities concerned with ensuring good standards of health and safety management

In order to assess and evaluate safety performance, and to introduce improvements which will reduce the potential for accidents, a number of safety monitoring techniques are used such as *auditing*, *inspections, sampling*, and *tours*

• SAFETY PHRASE

A safety phrase relating to dangerous substances listed in the approved supply list

In these Regulations, specific safety phrases may be designated by the letter 'S' followed by a distinguishing number or combination of numbers. The safety phrase must be quoted in full on any label or safety data sheet in which the safety phrase is required to be shown, e.g., S3 – keep in a cool place; S37 – wear suitable gloves, etc.

➔ *risk phrase* and *Approved Supply List*

• SAFETY POLICY

➔ *policy*

• SAFETY REPRESENTATIVE

An employee member of a trade union with at least two years' experience who is elected to represent his or her colleagues in discussions with the employer on health and safety matters

The employer must give the representative time off with pay for the purpose of carrying out the functions of a safety representative, and for training.

❑ ***The Safety Representatives and Safety Committees Regulations 1977***

• SAFETY SAMPLING

A technique designed to measure, by random sampling, the accident potential in a specific workshop or process by identifying safety defects or omissions

The area to be sampled is divided into sections and an observer appointed to each section. A prescribed route through the area is planned and observers follow their itinerary in a given time scale. During the sampling period safety defects are noted on a safety sampling sheet, with a limited number of points to be observed, e.g. housekeeping, eye protection, correct handling procedures, obstructed fire exits, faulty hand tools and damaged guards to machinery, and environmental factors such as lighting, noise and ventilation, etc.

Staff undertaking the inspections should be trained in the technique.

The results of the sampling activity can then be collated and presented in graphical form. The system monitors the effectiveness of the overall safety programme.

• SAFETY SIGNALS

Hand signals used in handling operations where the operator of plant, such as cranes, is unable to see the whole of the workplace

➔ *banksman; hand signal*

• SAFETY SIGN

A sign referring to a specific object, activity or situation

These provide information or instruction about health or safety at work by means of a signboard, a safety colour, an illuminated sign, an acoustic signal, a verbal communication or a hand signal.

❑ ***Health and Safety (Safety Signs and Signals) Regulations 1996***

• SAFETY SURVEYS

A detailed examination of a number of critical areas of operation or an in-depth study of the whole health and safety operation of premises

Following a survey, a report is published concerning the risks identified and the system for bringing about a gradual upgrading of standards. Recommendations would be phased according to the degree of risk, current legal requirements and the cost of eliminating or reducing the risks.

➔ *safety monitoring*

• SAFETY TOURS

Unscheduled examinations of a work area, carried out by a manager, possibly accompanied by health and safety committee members

This is to ensure, for example, that standards of housekeeping are at an acceptable level, fire protection measures are being observed and maintained, or personal protective equipment is being used correctly. To be effective, deficiencies noted during the tour should be remedied immediately.

➔ *safety monitoring*

• SAFETY TRAINING

Safety Training is a requirement under the HSWA 1974 and under each of the Six-Pack Regulations

'It shall be the duty of every employer to ensure, so far as is reasonably practicable, ... the matters to which that duty extends include in particular ... the provision of such information, instruction, training and supervision as is necessary to ensure the health and safety at work of all his employees'.

❑ ***Health and Safety at Work etc Act 1974***

• SALMONELLA

A type of bacterium which can cause infectious food poisoning

• SANITARY CONVENIENCES

Employers must provide suitable and sufficient sanitary conveniences and washing facilities at readily accessible places in workplaces

The rooms containing them must be adequately ventilated and lit; the rooms and conveniences must be kept clean and in an orderly condition; and separate rooms must be provided for men and women except where the convenience is in a separate room which can be locked from the inside.

❑ ***Workplace (Health, Safety and Welfare) Regulations 1992***

SCAFFOLD

A structure consisting of metal tubes and wooden boards to form a temporary working platform which is fixed to a building under construction to allow access for workmen, equipment and materials

SCAFFOLD BOARDS

Wooden boards used to form a working platform supported by scaffolding

The number of scaffold boards used depends on the purpose.

Purpose	Use of Platform	Number
Inspection and very light duty	Inspection, painting cleaning and access	3 boards
Light Duty	Plastering, stone cleaning, pointing and glazing	4 boards
General Purpose	General building brickwork, rendering and plastering	5 boards
Heavy Duty	Blockwork, brickwork and heavy cladding	5 boards
Masonry or special duty	Masonry, concrete, very heavy cladding	6-8 boards

• SCAFFOLDING

A scaffold is defined as: '... any temporarily provided structure on or from which persons perform work in connection with operations or works ... enables persons to obtain access to or which enables materials to be taken to any place at which such work is performed ...'

The legal requirements for the erection and use of scaffolding place obligations on contractors, anyone who controls work on a site, and employees/workers on a site.

Scaffolding must be erected by competent certificated persons and must be inspected before use, every week., after damage or alterations, and after very high winds.

❑ ***Construction (Health, Safety and Welfare) Regulations 1996.***

• SCIENTIFIC RESEARCH AND DEVELOPMENT

Scientific experimentation, analysis or chemical research carried out under controlled conditions including the determination of intrinsic properties, performance and efficacy as well as scientific investigation relating to product development

• SEALED SOURCE

A source containing any radioactive substance whose structure will prevent, under normal conditions of use, any dispersion of radioactive substances into the environment

It does not include any radioactive substance inside a nuclear reactor or any nuclear fuel element

❑ ***Ionising Radiations Regulations 1999***

• SEATING

Seating should be provided for employees wherever the work or task can be carried out whilst sitting

The seating should be appropriate for the task in hand

• SENSITISATION

A condition caused by initial exposure to a sensitising agent such as an allergen

Subsequent exposure triggers a reaction which can cause severe symptoms such as dermatitis or respiratory problems.

➔ *allergen, allergy*

• SERIOUS AND IMMINENT DANGER

Serious and Imminent Danger planning is required under the MHSW Regulations whereby an employer must establish, and carry out where necessary, appropriate procedures in the event of serious and imminent danger to employees and other persons resulting from his undertaking

• SEVERITY OF RISK

One of the factors used in risk assessment in order to arrive at a risk rating

• SEVERITY RATE

A procedure derived from accident statistics to calculate the number of working days lost through the number of employees at risk in a particular industry

The formula is:

$$\frac{\text{Total number of days lost} \times 1{,}000}{\text{Number of employees at risk}}$$

• SHORT-LIVED DAUGHTERS OF RADON 222

This curious phrase means polonium 218, lead 214, bismuth 214 and polonium 214

❑ ***Ionising Radiations Regulations 1999***

• SHORT-TERM EXPOSURE LEVEL (STEL)

The maximum concentration of airborne contamination to which a worker should not be exposed for more than a continuous period of 15 minutes.

➔ *maximum exposure level*

• SICK-BUILDING SYNDROME

A physical condition experienced by employees working in a particular premises or offices

Symptoms vary but include headaches, tiredness, nasal irritation, etc. The employee often does not make the connection between the symptom and the workplace. The causes are not clear but poor air quality is considered to be a major factor.

• SIEVERT

The unit of measurement of the biological effect of radiation on the human body

It is defined as a dose of radiation which delivers a joule of energy per kilogram of the mass of the recipient body. One Sievert is equal to 100 rem.

➔ *Rem*

• SIGNBOARD

A sign which provides information or instruction by a combination of geometric shape, colour and a symbol or pictogram and which is rendered visible by lighting of sufficient intensity

❑ ***Health and Safety (Safety Signs and Signals) Regulations 1996***

• SINGLE EUROPEAN ACT

This was an international agreement among the Member States of the Community which, amongst other things, introduced qualified majority voting on certain issues and the co-operation procedure

• SIX-PACK

A collective term for the 1992 European-derived legislation all of which emanated from European Directives

These were:

❑ ***Management of Health and Safety at Work Regulations 1992***
❑ ***Manual Handling Operations Regulations 1992***
❑ ***Display Screen Equipment Regulations 1992***
❑ ***Personal Protective Equipment at Work Regulations 1992***
❑ ***Provision and Use of Work Equipment Regulations 1992***
❑ ***Workplace (Health, Safety and Welfare) Regulations 1992***

• SKYLIGHTS

Usually, window openings in roofs or at high levels, unreachable by hand

Skylights must be capable of being opened, closed or adjusted in such a way that the person performing the operation is not exposed to any risk to health and safety

❑ ***Workplace (Health, Safety and Welfare) Regulations 1992***

• SLINGER

A person who attaches the load, which is to be lifted by a crane, to the crane's hook

The slinger must be trained in the use of the appropriate hand signals.

• SMOKE DETECTOR

Part of an automatic fire detection system

➔ *heat detector*

The detectors contain one or more sensors which react to smoke in the vicinity. They are usually sited at ceiling level in premises.

• SMOKING

Smoking in the workplace is a source of considerable concern for management and employees

Apart from being a fire hazard, there is an ongoing debate about the cancer risk from passive smoking. Under the Workplace Regulations, employers must make suitable arrangements for the protection of non-smokers from discomfort caused by tobacco smoke.

➔ *passive smoking*

❑ ***Workplace (Health, Safety and Welfare) Regulations 1992***
❑ ***HSE INDG63 Passive Smoking at Work***

• SO FAR AS IS REASONABLY PRACTICABLE

A phrase which occurs regularly in health and safety legislation

For example, *'It shall be the duty of every employer to ensure, so far as is reasonably practicable, the health, safety and welfare of all his employees.'*

It has been defined as follows:

'Reasonably practicable is a narrower term than physically possible and implies that a computation must be made in which the quantum of risk is placed in one scale and the sacrifice whether in time, money or trouble, involved in the measures necessary to avert the risk is placed in the other; and that, if it be shown there is a gross disproportion between them, the risk being insignificant in relation to the sacrifice, the person upon whom the duty is laid discharges the burden of proving that compliance was not reasonably practicable. This computation falls to be made at a point in time anterior to the happening of the incident complained of.'

📖 **Edwards v National Coal Board 1949**

This has been interpreted by the courts to mean that in order to carry out any duty which is subject to this proviso, the degree of risk involved has to be weighed against the time, trouble and cost in implementing the necessary measures to avoid the risk. If the cost of the measures is greatly disproportionate to the risk, there is no obligation to implement them.

• SO FAR AS IS PRACTICABLE

A regularly occurring phrase in health and safety legislation

It is a stricter standard than 'so far as is reasonably practicable' and has been interpreted by the courts to mean that whatever is technically possible in the light of current knowledge must be carried out.

📖 **Adsett v K & L Steelfounders & Engineers Ltd 1953**

• SOLE PLATE

A length of wood which is placed on the ground as a support to scaffolding

➔ *base plate* and *scaffold*

• SOLVENT

A dissolving agent

It can enter the body through various routes. Some solvents contain toxic ingredients which in contact with the body produce irritant, narcotic and toxic effects on the nervous system. ➔ *routes of entry*

• SPACE

Space in workplaces for each employee should be at least 11 cubic meters

In calculating the space no account need be taken of height exceeding 3 metres. ➔ *room dimensions*

❑ ***Workplace (Health, Safety and Welfare) Regulations 1992.***

• SPAD

The acronym used in railway terminology for 'signals passed at danger'

• SPALLING

The breaking up of the surface of a solid as result of exposure to extreme cold, corrosion, or intense heat followed by rapid cooling

• SPECIAL WASTE

Waste which is hazardous to deal with and whose disposal is subject to stringent regulations. Any controlled waste is special waste. Any other controlled waste, other than household waste is special waste.

❑ ***Special Waste Regulations 1996***

• SPORES

Cells formed from bacteria and moulds which are able to withstand adverse conditions including drying and heat

• SPRINKLER SYSTEM

A system of water pipes fitted with sprinkler heads, normally situated just below ceiling level and suitably spaced throughout an area, which is designed to extinguish a fire by the discharge of water

The water in the sprinkler head is held back by a glass bulb which shatters when a specified temperature is reached thus releasing the water.

➔ *quartzoid bulb*

• STACK EFFECT

The normal movement of air in a building due to difference between internal and external temperatures

Hot air inside the building rises and cool air is drawn in from outside. Where smoke is involved, this causes the hot smoke to rise within the building

• STAIN TUBE

A glass tube containing a chemically-treated adsorbent substance which partially changes colour when a quantity of air containing a gas or vapour is drawn through the tube

The length of stain indicates the degree or concentration of the contamination present in the sample. ➔ *grab sampling*

• STANDARD

1. A vertical tube used in scaffolding

➔ *scaffold*

2. A technical specification approved by a recognised standardising body for repeated or continuous application, with

which compliance is not compulsory; and, for the avoidance of doubt, includes a *harmonised standard* or a transposed harmonised standard

❑ ***Pressure Equipment Regulations 1999***

• STANDARD OF PROOF

The standard applied to establishing a person's guilt in court proceedings

In criminal law the standard is 'beyond a reasonable doubt'; in civil law the standard is based on 'the balance of probabilities'.

• STANDARD TEMPERATURE AND PRESSURE

S.t.p. are the normal standards relating to the measurement of gases, i.e. the pressure of a column of 760mm of mercury and the temperature of 0° Centigrade

• STATIC ELECTRICITY

An electrical 'charge' caused by friction when two materials or substances are rubbed together or come into contact with each other

• STATUTE

An Act of Parliament

• STATUTE LAW

A division of English law

It is written law produced through the parliamentary process. Statutes are Parliamentary legislation which supersede all other forms of law. Only Parliament can make, modify revoke or amend statutes.

• STATUTORY DUTY

An obligation imposed by a statute

• STATUTORY INSTRUMENT

A form of legislation which has been made by way of delegation from Parliament without going through the normal Parliamentary process

➔ *subordinate legislation*

• STEREOTYPING

The application of one's beliefs about a particular class or group of people to an individual member of that class or group which may or may not be accurate

• STERILISATION

A process whereby micro-organisms and viruses are killed by the application of heat or chemicals.

Spores can be killed by this process.

• STRESS

An adverse psychological, physiological reaction to a number of physical and emotional external stimuli

Continued exposure to stressful situations can lead to severe anxiety and depression.

📖 ***Walker v Northumberland County Council 1994***

• STROBOSCOBIC MOTION

The phenomenon by which static images viewed in rapid succession have the appearance of continuous movement

Conversely, rapidly revolving wheels can appear to be stationary especially in fluorescent lighting conditions which is dangerous in a work situation.

• STRUCTURAL ELEMENT

The structural elements of a building have been defined by the Building Regulations

These are:

- any member forming part of the frame of the building or any other beam or column not forming part of the roof structure
- floor, other than the lowest floor of the building
- an external wall
- a separating wall
- a compartment wall
- a structure enclosing a protected shaft
- a load-bearing wall
- a gallery

• STRUCTURE

Any building, steel or reinforced concrete structure (not being a building), railway line or siding, tramway line, dock, harbour,

inland navigation, tunnel, shaft, bridge, viaduct, waterworks, reservoir, pipe or pipeline (whatever, in either case it contains or is intended to contain), cable, aqueduct, sewer, sewage works, gas holder, road, airfield, sea defence works, river works, drainage works, earthworks, lagoon, dam, wall, caisson, mast, tower, pylon, underground tank, earth retaining structure or structure designed to preserve or alter any natural feature, and any other structure similar to the foregoing; or any framework, false work, scaffold or other structure designed or used to provide support or means of access during construction work; or any fixed plan in respect of work which is installation, commissioning, decommission-ing or dismantling and where such work involves a risk of a person falling more than 2 metres

❑ ***Construction (Design and Management) Regulations 1994***

• SUB-CONTRACTOR

A sub-contractor, under the CDM Regulations, is a contractor who is carrying out a specific part of a construction project

The sub-contractor must co-operate with the *principal contractor* and provide him or her with all relevant information relating to health and safety and pertaining to the risks associated with their work including any necessary control measures.

❑ ***Construction (Design and Management) Regulations 1994***

• SUBORDINATE LEGISLATION

Parliament has the power to delegate to Ministers, government departments and local authorities the power to make law relating to local matters such as the environment, planning and bye-laws

This type of law is termed 'subordinate legislation' and takes the form of Orders in Council, orders, rules, regulations, bye-laws and other instruments, all of which are collectively termed statutory instruments.

• SUBSTANCE

A chemical element or compound in the natural state or obtained by any production process, including any additive necessary to preserve the stability of the product and any impurity deriving from the process used

❑ ***Notification of New Substances Regulations 1993***

• SUITABLE HEAD PROTECTION

Head protection which is designed to provide protection, so far as is reasonably practicable, against foreseeable risks of injury to the head to which the wearer may be exposed; after any necessary adjustment, fits the wearer; and is suitable having regard to the work or activity in which the wearer may be engaged

❑ ***Construction (Head Protection) Regulations 1989***

• SUITABLE PERSON

A person appointed under the First-Aid Regulations to be in charge of a first-aid room, where one is provided, on work premises

➔ *first-aider* and *appointed person*

❑ ***Health and Safety (First-Aid) Regulations 1981***

• SUMMARY OFFENCE

An offence under the criminal law which is dealt with at magistrates' court

• SUPPLIER

A person or organisation which provides goods or services to another

A term met in a number of H & S regulations – e.g. CHIP, Provision and Use of Work Equipment and Supply of Machinery. The supplier has many obligations and requirements to ensure that anything he supplies satisfies regulations in relation to labelling and safety information.

❑ ***Chemicals (Hazard Information and Packaging for Supply) Regulations 1994***
❑ ***Provision and Use of Work Equipment Regulations 1998***
❑ ***Supply of Machinery (Safety) Regulations 1992***

• SUSPENDED ACCESS EQUIPMENT

➔ *boatswain's chair*

• SYMBIOSIS

The co-operation between organisms of different species which can lead to their mutual benefit

• SYMBOL OR PICTOGRAM

A figure which describes a situation or prescribes behaviour and which is used on a signboard or illuminated surface

❑ *Health and Safety (Safety Signs and Signals) Regulations 1996*

• SYNDROME

A group of symptoms occurring together which constitute a particular disorder

• SYNERGY

The interaction of two or more agents to produce a combined result greater than the sum of their individual effects

• SYNTHESIS

The combination of components to make a connected whole substance or compound

• SYSTEM

System, in electrical terms, means an electrical system in which all the electrical equipment is, or may be, electrically connected to a common source of electrical energy

• SYSTEM FAILURE

The unintentional release of stored energy (other than from a pressure relief system) from a pressure system

• TARGET ORGANS

Organs of the body which are particularly susceptible to attack from specific hazards

For example,

hazard:	dermatitis	target organ:	skin;
hazard:	irritant	target organ:	eyes, lung, skin.

• TECHNICAL FILE

A document kept by a manufacturer of machinery for a minimum period of 10 years after its manufacture

The file contains, amongst other things, technical drawings of the machinery and its circuits; design parameters, technical reports and test results; and a copy of the instructions for use of the machinery.

❑ ***Supply of Machinery (Safety) Regulations 1992***

• TEMPERATURE

How hot or cold the ambient working conditions are

The temperature in the workplace should be reasonable. What is reasonable is not defined in the Regulations but the accompanying Code of Practice recommends a minimum of 16°C, and where physical work is being carried out, the temperature may be 13°C. There is no recommended maximum temperature. ➔ *thermometer*

❑ ***Workplace (Health, Safety and Welfare) Regulations 1992.)***

• TENOSYNOVITIS

A condition, usually affecting the hands or wrists, caused by inflammation of the lining of the sheath which surrounds a tendon

It can be caused by working in an awkward position to carry out a task involving repetitive movements. ➔ *work-related upper limb disorder*

• TERATOGEN

An agent which can cause physical abnormalities in a developing embryo or foetus

• TETANUS

A disease of the central nervous system caused by infection of a wound by bacteria found in the soil and manure

• THERMOMETER

Thermometers must be provided in the workplace to enable employees to check the temperature

❑ *Workplace (Health, Safety and Welfare) Regulations 1992*

• THERMOPLASTIC

A thermoplastic material is one which becomes soft under the application of heat and which hardens when cool

• THERMO-SETTING

Relates to a material which is moulded into shape by heat and pressure

When subjected again to heat, the material is heat resistant up to the point of charring. Thermo-setting plastic is used to provide insulation for electrical circuits in many situations.

• THRESHOLD LIMIT VALUES

TLVs are the American terminology used to describe the limits for exposure to airborne concentrations of hazardous substances similar to the UK system of *occupational exposure limits*

• THRESHOLD SHIFT

A change in a person's hearing ability

It may be temporary as a result of sudden exposure to loud noise or permanent as a result of continued exposure to noise causing irreparable damage to the ear mechanism.

• THROUGH TIE

A scaffold tube used to connect a scaffold to a building

➔ *reveal tie*

A scaffold tube is placed vertically inside an opening or window in a building. and the through tie from the scaffold is coupled to it thus providing stability and support.

• TIME WEIGHTED AVERAGE

The exposure limit for a normal eight-hour day to which it is believed that workers may be exposed to a substance or condition day after day without adverse effect

➔ *maximum exposure limits*

• TINNITUS

A ringing noise heard in the ear

The noise, which is associated with hearing loss, is produced within the ear itself and is not associated with any external noise. ➔ *presbycusis*

• TOE BOARDS

150mm wide boards fitted at right angles to edge of scaffold boards to prevent materials from falling from the scaffold

• TOILET FACILITIES

Toilet facilities must be provided in the workplace for employees.

Since 1993, toilet facilities must be provided on the following scale:

1 – 5 persons	1 WC + 1 wash basins
6 – 25 persons	2 WC + 2 wash basins
26 – 50 persons	3 WC + 3 wash basins
51 – 75 persons	4 WC + 4 wash basins
76 – 100 persons	5 WC + 5 wash basins

❑ ***Workplace (Health, Safety and Welfare) Regulations 1992***

• TOOL-BOX TALK

An informal talk, discussion, or instruction which takes place on site in the workplace – usually given by a supervisor

Its content may be about the introduction or cessation of a particular method or system of work or any matter relevant to the immediate working situation.

• TOOLS

➔ *work equipment*

• TORT

An unlawful or wrongful act which causes harm to a person whether intended or not

It is a civil wrong which gives rise to a civil action for unliquidated damages. Among the principal torts are trespass, malicious damage, negligence and defamation.

• TORTFEASOR

A person who commits a tort

• TOTAL LOSS CONTROL

A management system aimed at preventing loss of any kind in an organisation by applying cost benefit analysis to all aspects of management including health and safety

• TOTAL INHALABLE DUST

Airborne material which is capable of entering the nose and mouth during breathing and is therefore available for deposition in the respiratory tract

➔ *respirable dust*

- **TOXIC**

Pertaining to poison

- **TOXIC AGENT**

A poisonous substance or liquid

- **TOXICITY**

A word used to quantify the effects of a toxic agent

- **TOXICOLOGY**

The study of poisons

- **TOXIN**

Any substance which is poisonous to an organism

It is a substance which, when absorbed by the body, has the potential to cause harm by affecting the body's metabolism.

- **TRADE UNION**

An organisation which consists wholly or mainly of workers of one or more descriptions whose principal purposes include the regulation of relations between workers and employers or employers' associations; or constituent or affiliated organisations which fulfill these conditions; or representatives of such constituent or affiliated organisations

An 'Independent Trade Union' means a trade union which is not under the domination or control of an employer, group of employers, or employers' associations and is not subject to interference from any such group arising out of the provision of financial or material support tending towards such control.

To be a recognised trade union, it must be on the list held by the Certification Officer and have applied for, and received, the Certificate of Independence from him and must be acknowledged by the employer for negotiating purposes.

- **TRADE UNION REPRESENTATIVE**

A person who is a member of a trade union who has been elected to represent the views of his or her membership in negotiations with the management of the organisation

• TRAFFIC ROUTE

Any route the purpose of which is to permit the access to or egress from any part of a construction site for any pedestrians or vehicles, or both, and includes any doorway, gateway, loading bay or ramp

❑ ***Construction (Health, Safety and Welfare) Regulations 1996***

• TRAINEE

A person undergoing training as part of his or her employment

Such people are covered by the provisions of HSWA.

• TRAINING

Training in health and safety has been a longstanding requirement of health and safety legislation

The requirement is repeated in all new legislation since 1992 but basic requirements are that employees must be trained on joining the organisation; on being exposed to new risks; on changing jobs; on the introduction of new techniques, new technology, or new systems of work. All training must be undertaken during company time and at company expense.

• TRAINING NEEDS ANALYSIS

An evaluation of an organisation's training requirements

The analysis follows a simple sequence of events:

- recognising the need for training
- identifying the type of training required
- deciding who or which departments require the training
- establishing the level of training required – managers – shop-floor – supervisors etc
- deciding on in-house or external trainers to carry out the training.

• TRANSFORMER

An electrical device which changes the voltage of an alternating current

The transformer is used to reduce the high voltages transmitted by power stations to the lower voltages used to operate domestic and some industrial electrical equipment.

• TRANSOM

A transverse horizontal metal tube fixed to two ledgers

➔ *ledger* and *scaffold*

• TRANSPORTABLE GAS CONTAINER

'Transportable gas containers' are defined as containers plus their permanent fittings of between 0.5 litres and 3,000 litres whose main purpose is for transporting the contents

Regulations cover the design, filling, examination, modification, repair, re-rating and the keeping of records.

❑ ***Carriage of Dangerous Goods (Classification, Packaging and Labelling) and Use of Transportable Pressure Receptacles Regulations 1996***

• TRANSPOSED HARMONISED STANDARD

A national standard of a member State which transposes a *harmonised standard*

❑ ***Pressure Equipment Regulations 1999***

• TRAUMA

A serious injury or a severe emotional shock

• TRAVELATORS

➔ *escalators*

• TRAVEL DISTANCE

The distance which can be travelled from any point in a building to a protected escape route, external escape route, or final exit

• TREATY OF ROME

The Treaty of Rome 1957 embraced two further treaties – the European Economic Community Treaty (EEC) and the European Atomic Energy Treaty

The EEC Treaty was the more important as it established the Common Market.

• TREATY ON EUROPEAN UNION

The Treaty on European Union 1992, also known as the

Maastricht Treaty, had the intention of transforming the Community from a basically economic organisation into a fully integrated European Union

Those objectives have not yet been fully realised because of the UK's reservations about monetary and social policies. The expression European Union (EU) became more commonplace as a descriptive term for the Community from this time.

• TREMCARD

The abbreviation for 'transport emergency card'

It is a small card carried by drivers of dangerous goods and members of emergency services containing coded information about hazardous substances to enable them to respond quickly to emergencies involving the transport of hazardous materials. ➔ *Appendix 5*

❑ ***Carriage of Dangerous Goods by Road Regulations 1996***

• TRESPASS

A voluntary wrongful act – the unlawful entry onto another's land or property

It is a civil wrong or tort.

• TRESPASSER

A person who enters land without invitation or permission

Although a trespasser is on land or on premises without the owner's or occupier's knowledge or permission, the occupier or proprietor owes him or her a duty of care.

📖 ***British Railways Board v Herrington 1972***

• TRIBUNAL

A legally constituted court which is empowered to hear cases under a number of Acts

Employment (formerly called industrial) Tribunals consist of a legally qualified chairman and two lay members. One lay member represents management and the other represents employees. These persons are usually from representative bodies, i.e. trade unions and management representative organisations. They are selected from a list kept by the Department of the Employment.

Parties appearing before tribunals are entitled to represent themselves or have a friend appear on their behalf. However, in the present climate of very complex employment legislation, it is common practice to employ lawyers.

Tribunals are empowered to hear cases under a number of Acts;

- ❑ ***Equal Pay Act***
- ❑ ***Trade Union and Labour Relations Act***
- ❑ ***Sex Discrimination Act***
- ❑ ***Race Relations Act***
- ❑ ***Employment Act***
- ❑ ***Health and Safety at Work etc Act***

• TRICHLOROETHYLENE

A highly hazardous, carcinogenic chemical substance, toxic by contact, inhalation and ingestion

It is used extensively as a cleaning agent and in a number of manufacturing processes.

• TRINITROTOLUENE

The high explosive commonly known as TNT

• TRIP DEVICE

Devices used in machine guarding to ensure that if the operator is likely to be injured, the devices will automatically stop the machine

They include pressure mats, emergency stop buttons and photo-electric systems.

• TRUCKS

➔ *fork-lift* and *reach truck*

• TYNDALL BEAM

A beam of light used to show up a pattern of dust particles in the atmosphere

• TWO-HAND CONTROL

A machine-guarding arrangement whereby it is not possible to operate the machine with one hand

On some guillotine machines, two buttons need to be pressed simultaneously to operate the machine. The buttons are spaced too far apart to allow them to be operated with one hand.

• ULTRAVIOLET RADIATION

Part of the electro-magnetic spectrum

As part of solar radiation, exposure to it can cause tissue damage (sunburn) leading to skin cancer and eye damage.

• UNDERGROUND SERVICES

Those services supplied by public utilities and local authorities, such as water, gas, electricity, sewage, and telephone

Their presence and direction can be identified by means of a pipe/cable detector (note – but not plastic) ➔ *excavations*

• UNLIQUIDATED DAMAGES

Unliquidated damages occur in tort cases where the amount of damages payable to an injured person are determined by a jury

• UNSAFE ACTS

Acts based on personal failures or omissions which may or may not lead to injury, damage or loss

• UNSAFE CONDITIONS

Conditions based on mechanical, physical or environmental faults, failures or breakdowns which may or may not lead to injury, damage or loss

• VAPOUR

The gaseous form of a solid or liquid affected by a rise in temperature which can cause solids and liquids to vaporise

• VENTILATION

The replacement of used air in an environment by fresh air using natural or mechanical means

Effective and suitable ventilation must be provided to ensure that there is a sufficient quantity of fresh or purified air in the workplace.

❑ ***Workplace (Health, Safety and Welfare) Regulation 1992***

• VERBAL COMMUNICATION

A predetermined spoken message communicated by a human or artificial voice

❑ ***Health and Safety (Safety Signs and Signals) Regulations 1996***

• VERMICULITE

A light, water-absorbent material formed from silicates which expand and exfoliate when heated

It is used in heat and sound insulating.

• VERMIN

Mammals, birds, worms and insects which carry disease

• VIBRATION WHITE FINGER

A physical condition caused by repeated and continual use of hand-held vibrating tools such as pneumatic hammers and drills and rotary hand tools

It affects the blood circulation in the hand and fingers which can cause gangrene. It is a *prescribed occupational disease*.

• VICARIOUS LIABILITY

➔ *liability*

• VIOLENCE

Violence to staff is 'any incident in which an employee is abused, threatened or assaulted by a member of the public in circumstances arising out of the course of his or her employment'.

Violence to staff is reportable under RIDDOR

- ❑ ***Reporting of Injuries, Diseases and Dangerous Occurrences Regulations 1995***
- ❑ ***IND(G)69L***

• VIRUS

A very small infectious and parasitic agent which grows within the cells of bacteria, animals and plants

Diseases caused by viruses include the common cold, influenza, measles, mumps, polio, and AIDS

• VISUAL DISPLAY UNIT

➔ *display screen equipment*

• VOLENTI NON FIT INJURIA

A Latin phrase meaning 'to a person who is willing no injury is done'

It is used in liability cases for injury or damages. In other words, the injured person has voluntarily accepted the risk of the harm that occurred and has consented to waive the duty of care owed to him by the defendant. For example, participants in a boxing match have freely consented to the risk of sustaining injury during the bout.

• VOLTAGE

Electrical potential energy is created by the separation of negative and positive charges

This electromotive force is measured in volts or voltage.

• VOTING

➔ *qualified majority vote*

• WARNING SIGN

A sign giving a warning of a risk to health or safety

❑ ***Health and Safety (Safety Signs and Signals) Regulations 1996***

• WASHING FACILITIES

'Facilities' means sanitary and washing facilities; 'sanitary accommodation' means a room containing one or more sanitary conveniences; and 'washing station' means a wash-basin or a section of a trough or fountain sufficient for one person

Where work activities result in heavy soiling of face, hands and forearms, the number of washing stations should be increased to one for every 10 people at work (or fraction of 10) up to 50 people; and one extra for every additional 20 people (or fraction of 20).

Where facilities provided for workers are also used by members of the public, the number of conveniences and washing stations specified above should be increased as necessary to ensure that workers can use the facilities without undue delay. ➔ *toilet facilities*

❑ ***Workplace (Health, Safety and Welfare) Regulations 1992***

• WASTE

➔ *controlled waste*

• WATCH

The acronym for the Working Group on the Assessment of Toxic Chemicals

• WATER

An adequate supply of drinking water must be provided in the workplace

❑ ***Workplace (Health, Safety and Welfare) Regulations 1992***

• WEIL'S DISEASE

Weil's Disease occurs in persons who come in contact with rats' urine

Persons most at risk are sewer workers. It is a reportable disease under *RIDDOR*. ➔ *leptospirosis*

• WELFARE FACILITIES

Welfare facilities include toilets, canteens, rest rooms, drinking water, first-aid, etc.

❑ ***Workplace (Health, Safety and Welfare) Regulations 1992***
❑ ***Construction (Health, Safety and Welfare) Regulations 1996***

• WHISTLE BLOWING

A term used to describe the conduct of an employee who reports misconduct by an employer which is against the public interest

• WHITE PAPER

A government consultation document following on from the publication of a *green paper*

It contains firm proposals for incorporation in legislation. It is a further stage in the development of parliamentary legislation

• WINDOWS

Windows must be capable of being safely opened and cleaned. When in the open position, they must not cause a danger to any persons

❑ ***Workplace (Health, Safety and Welfare) Regulations 1992***

• WOMEN

➔ *maternity*

• WOOLF REPORT

The result of an inquiry by Lord Woolf in 1994-6 into costs, delays and other matters associated with the administration of civil justice

His recommendations included, amongst others, making compensation easier to implement; the introduction of a 'fast-track' system for dealing with personal injury claims; and a change in the expert witness procedure whereby each side use the same expert witness whose written report is available to both sides before going to court.

❑ ***Civil Procedure Rules 1998***

• WORKER

An individual who works under (or, worked under) a contract of employment; or any other contract, whether oral or in writing, whereby the individual undertakes to do or perform personally any work or services for another party to the contract whose status is not that of a client or customer of any profession or business undertaking carried on by the individual and any reference to a worker's contract shall be construed accordingly

❑ ***Working Time Regulations 1998***

• WORK EQUIPMENT

Any machinery, appliance, apparatus, tool or installation for use at work whether exclusively or not

According to the Work Equipment Code of Practice, the term includes:

- 'tool box tools' – hammers, knives, handsaws, meat cleavers etc
- single machines such as – drilling machines, circular saws,
- photocopiers, combine harvesters, dumper trucks, etc
- apparatus such as – laboratory apparatus (Bunsen burners, etc)
- lifting equipment such as hoists, lift trucks, elevating work platforms, lifting slings, etc., and
- other equipment such as ladders, pressure water cleaners, etc

❑ ***Provision and Use of Work Equipment Regulations 1998***

• WORKING PLATFORM

Any platform used as a place of work or for access to a place of work

This will include a scaffold, suspended scaffold, cradle, mobile platform, trestle, gangway, run, gantry, stairway, and crawling ladder

❑ ***Construction (Design and Management) Regulations 1994***

• WORKING TIME

Any period during which a worker is working, at his employer's disposal and carrying out his activity or duties; any period during which he is receiving relevant training, and any additional period which is to be treated as working time under a relevant agreement and 'work' shall be construed accordingly

❑ ***Working Time Regulations 1998***

• WORKPLACE

1 'Any premises which are not domestic premises and are made available to any person as a place of work, and includes any place within the premises to which such person has access while at work; any room, lobby, corridor, staircase, road or other place used as a means of access to or egress from the workplace or where facilities are provided for use in connection with the workplace, other than a public road

❑ ***Workplace (Health, Safety and Welfare) Regulations 1996***

2 Workplace means, in relation to an employee, any place or places where that employee is likely to work or which he is likely to frequent in the course of his employment or incidentally to it and, in relation to a representative of employee safety, any place or places where the employees he represents are likely to so work or frequent

❑ ***Health and Safety (Consultation with Employees) Regulations 1996***

• WORK-RELATED UPPER LIMB DISORDER

A physical condition which may affect hands, arms, wrists, shoulders and neck

The symptoms are associated with the use of display screen equipment but they can arise in other occupations such as fruit-picking and poultry processing. ➔ *tenosynovitis*

• WORK SPACE

➔ *space*

• WORKSTATION

A ny furniture and equipment used by a worker together with the immediate environment in which the work takes place

❑ ***Workplace (Health, Safety and Welfare) Regulations 1992***

• WRIT

A document usually issued by a court ordering the person to whom it is addressed to do, or cease to do, some act

• X-RAY

An electro-magnetic radiation with considerable penetrating power

It is an ionising radiation which can cause tissue damage leading to cancer.

• YOUNG PERSONS

A young person in the context of health and safety is a person under 18 years of age

An employer shall not employ a young person unless he has made a risk assessment in relation to that young person.

The young person's risk assessment shall take account of:

- his or her inexperience, lack of awareness of risks, and immaturity
- the fitting-out and layout of the workplace and workstation
- the nature, degree and duration of exposure to physical, biological and chemical agents
- the form, range, and use of work equipment and the way in which it is handled
- the organisation of processes and activities
- the extent of health and safety training provided to the young person
- the risks from certain agents, processes and work.

For five or more employees, the assessment must record the significant findings and any group of employees identified as being particularly at risk.

❑ ***Management of Health and Safety at Work Regulations 1998***

• YOUNG WORKER

A worker who has attained the age of 15 but not the age of 18 and who, in England and Wales, is over compulsory school age

❑ ***Working Time Regulations 1998***

• ZONE

An area protected by a group of automatic fire detectors

When a detector initiates a signal to a central control point, the signal identifies the zone where the fire situation has occurred.

• ZOONOSE

A disease which can be transmitted from animals to humans

Ringworm and leptospirosis are examples of these.

❑ ***HS(G)105***

Appendix 1

Health and Safety Legislation

Factories Act 1961

Offices, Shops and Railway Premises Act 1963

Employer's Liability (Compulsory Insurance) Act 1969

Fire Precautions Act 1971

Health and Safety at Work etc Act 1974

Employment Protection Act 1975

Safety Representatives and Safety Committees Regulations 1977

Social Security (Industrial Diseases)(Prescribed Diseases) Regulations 1980

Health and Safety (First-Aid) Regulations 1981

Occupier's Liability Act 1957 and 1984

Ionising Radiations Regulations 1985 and 1999

Fire Safety and Safety of Places of Sport Act 1987

Control of Asbestos at Work Regulations 1987

Electricity at Work Regulations 1989

Noise at Work Regulations 1989

Food Safety Act 1990

Environmental Protection Act 1990

Management of Health and Safety at Work Regulations 1992 and 1999

Manual Handling Operations Regulations 1992

Health and Safety (Display Screen Equipment) Regulations 1992

Personal Protective Equipment at Work Regulations 1992

Provision and Use of Work Equipment Regulations 1992 and 1998

Workplace (Health, Safety and Welfare) Regulations 1992

Supply of Machinery (Safety) Regulations 1992

Construction (Design and Management) Regulations 1994

Chemicals (Hazard Information and Packaging for Supply) Regulations 1994

Disability Discrimination Act 1995

Reporting of Injuries, Diseases and Dangerous Occurrences Regulations 1995

Special Waste Regulations 1996

Construction (Health, Safety and Welfare) Regulations 1996

Health and Safety (Consultation with Employees) Regulations 1996

Health and Safety (Signs and Signals) Regulations 1996

Carriage of Dangerous Goods (Classification, Packaging and Labeling) and Use of Transportable Pressure Receptacles Regulations 1996

Fire Safety (Workplace) Regulations 1997

Health and Safety (Young Persons) Regulations 1997

Control of Major Accident Hazards Regulations 1999

Working Time Regulations 1998 and 1999

Pressure Equipment Regulations 1999

Control of Lead at Work Regulations 2002

Control of Substances Hazardous to Health Regulations 2002

NB. The above list is not exhaustive and legislation is constantly being revised, consolidated and revoked. The reader should consult HMSO for details of current legislation.

Appendix 2

Approved Codes of Practice

COP Series

Control of lead at work - approved code of practice	COP 2
Standards of training in safe gas installation - approved code of practice 1987	COP 20
Safety in Docks: Docks Regulations 1988 - approved code of practice	COP 25
Safety of exit from mines underground workings - approved code of practice 1988	COP 28
First-aid on offshore installations and pipeline works - approved code of practice with regulations and guidance 1990	COP 32
Safety of pressure systems: Pressure Systems and Transportable Gas Containers Regulations 1989 - approved code of practice 1989	COP 37

L Series

A Guide to the Health and Safety at Work etc Act 1974: guidance on the Act	L 1
Management of health and safety at work. Management of Health and Safety at Work Regulations 1999 - approved code of practice 2000	L 21
Workplace health, safety and welfare. Workplace (Health, safety and Welfare) Regulations 1992 - approved code of practice and guidance 1992	L 24

The control of asbestos at work: Control of Asbestos at Work Regulations 1987 - approved code of practice 1999 L 27

First aid at work: The Health and safety (First-Aid) Regulations 1981 - approved code of practice and guidance 1997 L 74

Work with ionising radiation: Ionising Radiations Regulations 1999 - approved code of practice L 121

NB The above lists are not exhaustive and the reader should consult HSE Books for up-to-date and current information on publications.

Appendix 3

HEALTH AND SAFETY GUIDANCE NOTES

Chemical Safety Series

Storage and use of sodium chlorate and other similar strong oxidants 1998	CS 3
Storage and use of LPG on metered estates 1987	CS 11
The cleaning and gas freeing of tanks containing flammable residues 1985	CS 15
Storage and handling of ammonium nitrate 1986	CS 18
Storage and handling of organic peroxides 1991	CS 21
Fumigation 1996	CS 22

Environmental Hygiene Series

Cadmium: health and safety precautions 1995	EH 1
Chromium and its inorganic compounds: health and safety precautions 1998	EH 2
Asbestos: exposure limits and measurement of airborne dust concentrations 1995	EH 10
Beryllium: health and safety precautions 1995	EH 13
Isocyanates: health hazards and precautionary measures 1999	EH 16
Mercury and its inorganic divalent compounds 1996	EH 17

General Series

Safety in pressure testing 1998	GS 4
Avoidance of danger from overhead electrical lines 1997	GS 6

Safe erection of structures: Part 2. Site management and procedures 1985	GS 28/2
Health and safety in shoe repair premises 1984	GS 32
Electrical test equipment for use by electricians 1995	GS 38
Pre-stressed concrete 1991	GS 49

Medical Series

Colour vision 1987	MS 7
Mercury: medical guidance notes 1996	MS 12
Asbestos: medical guidance notes 1996	MS 13
Biological monitoring of workers exposed to organo-phosphorous pesticides 2000	MS 17
Health surveillance of occupational skin disease 1998	MS 24
Medical aspects of occupational asthma 1998	MS 25

Plant and Machinery Series

High temperature textile dyeing machines 1997	PM 4
Automatically controlled steam and hot water boilers 1989	PM 5
Safety in the use of pallets	PM 15
Eyebolts 1978	PM 16
Pneumatic nailing and stapling guns 1979	PM 17
Safety at rack and pinion hoists 1981	PM 24

Legal Series (L) See Approved Codes of Practice - Appendix 2

NB The above lists are not exhaustive and the reader should consult HSE Books for details of up-to-date and current publications

Appendix 4

Health and Safety Guidance Booklets

Safety advice for bulk chlorine installations	HSG 28
Pie and tart machines	HSG 31
Anthrax	HSG 36
An introduction to local exhaust ventilation	HSG 37
Lighting at work	HSG 38
Compressed air safety	HSG 39
Safe handling of chlorine from drums and cylinders	HSG 40
Safety in the use of metal cutting guillotines and shears	HSG 42
Industrial robot safety	HSG 43
Reducing error and influencing behaviour	HSG 48
Storage of flammable liquids in containers	HSG 51
The maintenance, examination and testing of local exhaust ventilation	HSG 54
Seating at work	HSG 57
Health surveillance at work	HSG 61
Health and safety in tyre and exhaust fitting premises	HSG 62
Successful health and safety management	HSG 65
Protection of workers and the general public during the development of contaminated land	HSG 66
Health and safety in motor vehicle repair	HSG 67
Chemical warehousing	HSG 71
Control of respirable silica dust in heavy clay and refractory processes	HSG 72

Control of respirable crystalline silica in quarries	HSG 73
Health and safety in retail and wholesale warehouses	HSG 76
Dangerous goods in cargo transport units	HSG 78
Health and safety in golf course management and maintenance	HSG 79
Electricity at work	HSG 85
Safety in the remote diagnosis of manufacturing plant and equipment	HSG 87
Hand-arm vibration	HSG 88
Safeguarding agricultural machinery	HSG 89
The law on VDUs; An easy guide	HSG 90
Safe use and storage of cellular plastics	HSG 92
The assessment of pressure vessels operating at low temperature	HSG 93
Safety in the design and use of gamma and electron irradiation facilities	HSG 94
The radiation safety of lasers used for display purposes	HSG 95
The costs of accidents at work	HSG 96
A step by step guide to COSHH assessment	HSG 97
Prevention of violence to staff in banks and building societies	HSG 100
The costs to Britain of workplace accidents and work-related ill health in 1995/96	HSG 101
Safe handling of combustible dusts	HSG 103
Control of noise in quarries	HSG 109
Maintaining portable and transportable electrical equipment	HSG 107
Seven steps to successful substitution of hazardous substances	HSG 110
Seven steps to successful substitution of hazardous substances (in Welsh)	HSG 110W
Lift trucks in potentially flammable atmospheres	HSG 113
Conditions for the authorisation of explosives in Great Britain	HSG 114
Manual Handling	HSG 115

Making sense of NONS	HSG 117
Electrical safety in arc welding	HSG 118
Manual handling in drinks delivery	HSG 119
A pain in your workplace?	HSG 121
New and expectant mothers at work	HSG 122
Working together on firework displays	HSG 123
Giving your own firework display	HSG 124
A brief guide on COSHH for the offshore oil and gasIndustry	HSG 125
Energetic and spontaneously combustible substances	HSG 131
How to deal with sick building syndrome	HSG 132
Preventing violence to retail staff	HSG 133
Storage and handling of industrial nitrocellulose	HSG 135
Workplace transport safety	HSG 136
Health risk management	HSG 137
Sound solutions	HSG 138
Safe use of compressed gases in welding, flame cutting and allied processes	HSG 139
Safe use and handling of flammable liquids	HSG 140
Electrical safety on construction sites	HSG 141
Dealing with offshore emergencies	HSG 142
Dispensing petrol	HSG 146
Health and safety in construction	HSG 150
Protecting the public	HSG 151
Railway safety principles and Guidelines. Part 1	HSG 153/1
Railway safety principles and guidance. Part 2Section A	HSG 153/2
Railway safety principles and Guidelines. Part 2. Section C	HSG 153/4
Railway safety principles and guidance . Part 2Section D	HSG 153/5
Railway safety principles and guidance. Part 2Section E	HSG 153/6

Railway safety principles and guidance. Part 2.Section F	HSG 153/7
Railway safety principles and guidance. Part 2.Section G	HSG 153/8
Managing crowds safely	HSG 154
Slips and trips	HSG 155
Slips and trips	HSG 156
Flame arresters	HSG 158
The carriage of dangerous goods explained. Part 1	HSG 160
The carriage of dangerous goods explained. Part 2	HSG 161
The carriage of dangerous goods explained	HSG 162
The carriage of dangerous goods explained Part 3	HSG 163
The carriage of dangerous goods explained Part 5	HSG 164
Formula for health and safety	HSG 166
Biological monitoring in the workplace	HSG 167
Fire safety in construction work	HSG 168
Camera operations on location	HSG 169
Vibration solutions	HSG 170
Well handled	HSG 171
Health and safety in sawmilling	HSG 172
Monitoring strategies for toxic substances	HSG 173
Fairgrounds and amusement parks. Guidance on safe practice	HSG 175
Storage of flammable liquids in tanks	HSG 176
The spraying of flammable liquids	HSG 178
Application of electro-sensitive protective equipment using light curtains and light beam devices to machinery	HSG 180
Assessment principles for offshore safety cases	HSG 181
Sound solutions offshore	HSG 182
5 steps to risk assessment	HSG 183
Guidance on the handling, storage and transport of airbags and seatbelt pretensioners	HSG 184

The bulk transfer of dangerous liquids and gases between ship and shore — HSG 186

NB The above lists are not exhaustive and the reader should consult HSE Books for details of up-to-date and current publications

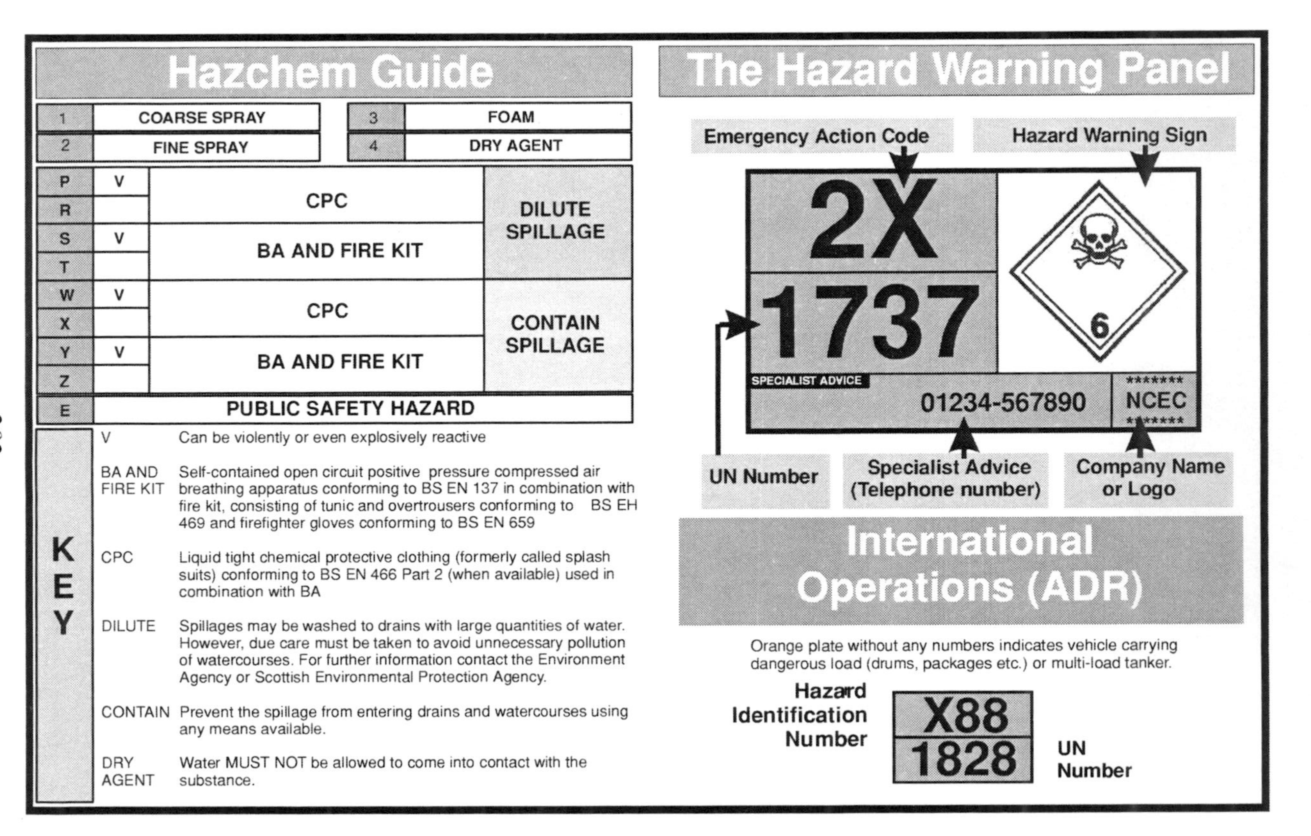
Hazchem Guide
1
COARSE SPRAY
2
FINE SPRAY
3
FOAM
4
DRY AGENT
P V
R
S V
T
W V
X
Y V
Z
CPC
BA AND FIRE KIT
CPC
BA AND FIRE KIT
DILUTE SPILLAGE
CONTAIN SPILLAGE
E
PUBLIC SAFETY HAZARD
KEY
V Can be violently or even explosively reactive
BA AND FIRE KIT Self-contained open circuit positive pressure compressed air breathing apparatus conforming to BS EN 137 in combination with fire kit, consisting of tunic and overtrousers conforming to BS EH 469 and firefighter gloves conforming to BS EN 659
CPC Liquid tight chemical protective clothing (formerly called splash suits) conforming to BS EN 466 Part 2 (when available) used in combination with BA
DILUTE Spillages may be washed to drains with large quantities of water. However, due care must be taken to avoid unnecessary pollution of watercourses. For further information contact the Environment Agency or Scottish Environmental Protection Agency.
CONTAIN Prevent the spillage from entering drains and watercourses using any means available.
DRY AGENT Water MUST NOT be allowed to come into contact with the substance.
The Hazard Warning Panel
Emergency Action Code
Hazard Warning Sign
2X
1737
6
SPECIALIST ADVICE
01234-567890
NCEC
UN Number
Specialist Advice (Telephone number)
Company Name or Logo
International Operations (ADR)
Orange plate without any numbers indicates vehicle carrying dangerous load (drums, packages etc.) or multi-load tanker.
Hazard Identification Number
X88
1828
UN Number

The hazard identification number consist of two or three figures. In general the figures indicate the following hazards:

2	Emission of gas due to pressure or to chemical reaction	5	Oxidising (fire-intensifying) effect
3	Flammability of liquids	6	Toxic or risk of infection
4	Flammability of solids or self-heating liquid	7	Radioactivity
		8	Corrosivity
		9	Risk of spontaneous violent reaction

Doubling of a figure indicates an intensification of that particular hazard.

If the HIN is prefixed with an 'X' this indicates the substance will react dangerously with water. For such substances, water may only be used by approval of experts.

Where the hazard associated with a substance can be adequately indicated by a single figure, this is followed by a zero.

The following combinations of figures, however, have a special meaning:

22	refrigerated liquified gas, asphyxiant
323	flammable liquid which reacts dangerously with water, emitting flammable gases
333	pyrophoric liquid
362	flammable liquid, toxic, which reacts with water, emitting flammable gases
382	flammable liquid, corrosive, which reacts with water emitting flammable gases
423	solid which reacts with water, emitting flammable gases
44	flammable solid, in the molten state at an elevated temperature
446	flammable solid, toxic, in the molten state at an elevated temperature
462	toxic solid which reacts with water, emitting flammable gases
482	corrosive solid which reacts with water, emitting flammable gases
539	flammable organic peroxide
606	infectious substance
623	toxic liquid, which reacts with water, emitting flammable gases
642	toxic solid, which reacts with water, emitting flammable gases
823	corrosive liquid which reacts with water, emitting flammable gases
842	corrosive solid which reacts with water, emitting flammable gases
90	environmentally hazardous substance, miscellaneous
99	miscellaneous dangerous substance carried at an elevated temperature

This information and the coloured panel on the back cover are reproduced with the kind permission of the National Chemical Emergency Centre, F6 Culham, Abingdon, Oxon., OX14 3ED. Website: www.the-ncec.com or contact via acec@aeat.co.uk.